Research & Development Aircraft No.2

X-32

The Boeing Joint Strike Fighter

Hugh Harkins

X-32, The Boeing Joint Strike Fighter

Published by Centurion Publishing
PO Box 3268
Glasgow
United Kingdom
G65 9YE

ISBN: 978-1-903630-20-4

This volume first published 2013

The Publisher would like to thank the following for their assistance and contribution in preparation of this publication: BAE SYSTEMS; The Boeing Company; General Electric Aircraft Engines; Lockheed Martin; Northrop Grumman; NASA; Pratt & Whitney; the USAF and USN.

X-32, The Boeing Joint Strike Fighter

Contents

The Boeing X-32 Concept Demonstrator Aircraft was the losing contender in the international Joint Strike Fighter program, which led to the Lockheed Martin F-35 Lightning II 5th generation strike fighter entering service with air forces on both sides of the Atlantic. This volume details the genesis of the Joint Strike Fighter program and describes the short take off and vertical landing and conventional landing variants of the X-32 design. The evolution of the X-32 into the preferred weapon system concept, which would have been the service models had fortune shone more brightly on the program, is covered, as is the development and flight test program. A chronology details the flight test program of the rival Lockheed Martin X-35. While the X-32 aircraft were retired to museums, the design heritage of which they were a part has been carried over to Boeings 6th generation fighter studies. These sixth generation fighter concepts emerging in the second decade of the 21st Century are aimed at a potential service entry sometime after 2025.

Chapter One

The Road to Joint Strike Fighter

What eventually emerged as the JSF program was a culmination of technology demonstrations and research programs initiated to find technologies for incorporation into replacements for a number of legacy aircraft on both sides of the Atlantic. The X-35A CDA seen above was the Lockheed Martin entry into the competition to provide a common fighter capable of meeting the requirements of the USAF, USMC, USN and the British RN and RAF. LMTAS

The Joint Strike Fighter (JSF) program was initiated to produce a common airframe, which offered conventional take-off and landing (CTOL), short take-off and vertical landing (STOVL) and carrier (CV) variants of an affordable multi-role combat aircraft. The winning contender would be required to replace a number of aircraft types in service on both sides of the Atlantic. These included the Lockheed Martin F-16 Fighting Falcon and Northrop Grumman A-10A Thunderbolt II in USAF service. The USN and USMC required replacements for the first generation Boeing F/A-18 Hornet and the Boeing/BAE Systems AV-8B Harrier II, while the UK required replacements for the BAE Systems Sea Harrier F/A2 in Royal Navy (RN) service and the Harrier GR.7/9 in Royal Air Force (RAF) service, these types being retired from service in 2006 and 2011 respectively.)

There was initially a US/UK domestic requirement for almost 3,000 JSF aircraft and potential for 1,000+ export orders, with many countries participating in the program at various levels. However, up to and including the Concept Demonstrator Phase, only the US and UK were full partners in the program and able to influence the design of the aircraft, although to what extent the UK was realistically able to influence the design remains unclear.

Boeing's entry into the competition to build the JSF was the X-32 CDA. Here the X-32B Short Take-Off and Vertical Landing variant is seen getting airborne after a short take-off run. Boeing

The need for a future Harrier replacement was highlighted by the 1982 Falklands conflict between Britain and Argentina. During this conflict, RN Sea Harrier FRS.1 and RAF Harrier GR.3's operated from the small British aircraft carriers *HMS Hermes* and *HMS Invincible,* providing crucial air defence for the fleet and ground attack capabilities, without which the British operation to retake the Falklands would have been all but impossible.

The United States Marine Corp (USMC) need for a replacement for its first generation AV-8C Harriers saw the development and service entry of the McDonnell Douglas/BAe (now Boeing and BAE Systems) AV-8B Harrier II second generation Harrier, the prototype of which flew for the first time in autumn 1981. The first British Harrier GR.5 flew for the first time on 30 April 1985 and the same year the first USMC AV-8B squadron was declared operational. The first RAF unit to receive production Harrier GR.5's was the Harrier Conversion Team (HCT), which reformed at RAF Wittering on 1 March 1987, although the GR.5 was not operational until March 1988.

USMC AV-8B's were deployed to the Persian Gulf and contributed to the 1991 Gulf War against Iraq, proving the types close air support capability. Later in the 1990's USMC AV-8B's and RAF Harriers were frequently deployed to the Gulf region and to the troubled Balkans. In 1999, USMC AV-8B's and RAF Harrier GR.7's participated in the 79 day operation Allied Force air campaign over Serbia and Kosovo. Although proving to be extremely capable in service the second generation Harrier had major shortcoming. Its lack of airborne radar meant that it was unable to operate with radar guided air to air missiles, such as the AIM-120 AMRAAM (Advanced Medium Range Air to Air Missile). This shortcoming was addressed by introduction of the Harrier II plus, which became the standard variant of the AV-8B from the 205th production single seat aircraft for the USMC, which was the first fully equipped combat capable Harrier II+.

The USMC had the most pressing need for the new aircraft, which emerged as the JSF, requiring the aircraft to replace its fleet of Boeing/BAE Systems AV-8B Harrier II/+ and Boeing F/A-18 Hornet strike fighters. In the 1990's, the USMC began to receive the radar equipped Harrier II+ seen above, but required the JSF to enable transition to an all-supersonic STOVL fighter force. Boeing

Flying for the first time on 22 September 1992, this was the first true prototype of the Harrier II Plus, although another AV-8B, 161397, the second Full Scale Development (FSD) aircraft had flown configured with a dummy nose cone as an aerodynamic test bed.

The main changes included the fitting of a Hughes (now Raytheon) AN/APG-65 radar as fitted in the FA-18 Hornet, which allowed the Harrier II to carry and fire the fully active AIM-120 AMRAAM, bestowing a beyond visual range (BVR) interception capability, as well as allowing the launch of the Boeing AGM-84 Harpoon anti ship missiles.

The Harrier II Plus, which saw the Harrier II metamorphous into a truly multi-role Short Take-Off and Vertical Landing (STOVL) strike fighter, was developed through a three-nation agreement among the United States, Spain and Italy. The final 27 aircraft for the USMC were built to Harrier II Plus standard, although the service only received 24 of the aircraft; the other three going to the Italian Navy. The first new production Harrier II Plus was delivered on 18 April 1993.

In May 1994, MDC (McDonnell Douglas Corporation) received a contract to upgrade 73 USMC AV-8Bs to Harrier II Plus standard, this total was reduced from the 114 that was originally planned. The conversion included a completely new fuselage, as this was less expensive than re-modelling the old one to accommodate the APG-65 radar and Forward Looking Infra-Red (FLIR).

The Spanish Navy received 12 EAV-8Bs and a single TAV-8B, the latter ordered in 1992, to replace the first generation AV-8S previously operated from the carrier *Dedalo,* now replaced by the *Principi de Austurias.* In November 1992, Spain ordered 13 Harrier II Plus with delivery commencing in early 1995 and the last aircraft being delivered in July 1997. This aircraft was one of eight that were manufactured by CASA in Spain. Later contracts were awarded for the re-manufacturing of Spain's EAV-8B day-attack Harriers into Harrier II Plus standard.

The fourth operator of the Harrier II was Italy, which ordered a pair of TAV-8B Harrier II's in May 1989. These aircraft were delivered in August 1991. In July 1991, sixteen Harrier II Plus aircraft were ordered, with the first three aircraft handed over on 20 April 1994. The last 13 aircraft were license built in Italy by Alenia, with the 13th and final aircraft delivered on 25 November 1997.

Even before the AV-8B Harrier II had entered service, Anglo-US thoughts were turning to their eventual replacement. In September 1983, representatives from the US and Britain met at that years Society of British Aerospace Companies (SBAC) show at Farnborough to formally establish an Advanced Short Take-Off and Vertical Landing (ASTOVL) program. The basics of the program were aimed at jointly developing technologies for a 'viable' single-engine ASTOVL fighter that could enter service around the year 2000. In January 1986, a Memorandum of Understanding (MoU) was signed by the governments of the US and UK outlining a collaborative program on research for future ASTOVL technologies required for a possible replacement for their respective fleets of AV-8B Harrier II and Harrier GR.5/7 STOVL strike fighters. A boost for the program came in 1987 when the USMC revealed that it would also consider the ASTOVL as a replacement for its McDonnell Douglas F/A-18 Hornet strike fighters.

On 1 February 1989, it was announced that NASA in co-operation with the US DoD (Department of Defence) and UK MoD (Ministry of Defence) had completed the concept evaluation studies of ASTOVL fighter concepts. The aim of the ASTOVL studies was to assess the overall propulsive-lift/airframe system performances and to identify the criticality of various technologies needed to attain these performances. NASA's Lewis Research Centre in Cleveland, Ohio led the propulsion system study efforts and NASA's Ames Research Centre in California led the airframe and integration efforts. British Aerospace and Rolls Royce conducted all the UK studies and the US studies were conducted by McDonnell Douglas in association with Pratt & Whitney, Grumman (now Northrop Grumman) with General Dynamics (now Lockheed Martin) with General Electric and Lockheed with Pratt & Whitney/Rolls-Royce. The concept evaluation studies resulted in two sets of studies each of four concepts:

Top: As well as replacing the USMC fleet of AV-8B STOVL fighters, the JSF was aimed at fielding a replacement for the Royal Navy Fleet Air Arm Sea Harrier FA.2's and RAF Harrier GR.7/9 ground attack aircraft. BAE Systems **Centre: The USAF planned to replace a number of types with the JSF, including the A-10A ground attack aircraft.** USAF **Above: While the USN is fielding over 500 Boeing F/A-18E/F/EA-18G strike-fighters and electronic attack aircraft it still required the JSF as a 'first day of war' strike aircraft as well as to replace older first generation Hornets.** USN

- Advanced Vectored Thrust (AVT): During the course of the studies it became apparent that one AVT variant - the Mixed Flow Vectored Thrust (MFVT) was sufficiently different to merit treatment as a separate concept.
- Remote Augmented Lift System (RALS): This system had an aft mounted engine along conventional lines with a vectoring nozzle. To enable STOVL operations air was ducted from the engine to an afterburner, which was mounted vertically in the forward fuselage.
- Ejector Augmenter (EA): This system would duct engine exhaust augmented by ambient air through full span ejector flaps. This system had been unsuccessfully tried in 1972 on the Rockwell XFV-12.
- Hybrid Tandem Fan (HTF): The HTF system would have a two-section fan mounted on a drive shaft. For STOVL operations valves diverted the front fan exhaust through vectoring nozzles, thus enabling the aircraft to achieve vertical lift. In normal cruise flight the engine operated as a conventional turbofan.

The results of the international concept evaluation studies were reviewed by an executive steering committee, which was to determine the overall direction for the future US/UK ASTOVL Technology Development Program. The overall results of the above activities indicated that the most promising ASTOVL configurations were those which utilised remote lift for jet-borne flight (de-coupling the location of the engine from the placement of the jet thrust nozzles) and conventional mixed flow propulsive systems for wing-borne flight. It was announced that these concepts would receive the greatest attention in the technology development program to facilitate possible future development of an ASTOVL combat aircraft. However, doubts over the workability of all the concepts and lack of funding for development saw the program simmer along with the threat of cancellation ever present.

In 1989, the Defence Advanced Research Projects Agency (DARPA later ARPA) asked the USN/USMC to prepare an operational capability document, a move which would change the focus of the ASTOVL study. This focused on cost and operability goals of the program, instead of the traditional focus on performance targets. The document stated a deck spot factor no greater than that of the F/A-18 and incorporation of either the General Electric YF120 or the Pratt & Whitney YF119 engines then being developed for the USAF ATF (Advanced Tactical Fighter) program. The high power available from these power plants opened up a whole new bag of tricks for the ASTOVL program.

In the early 1980s, USN planners were looking for a suitable replacement for the services premier strike aircraft, the Grumman A-6 Intruder II, which had made its first flight in prototype A2F-1 form as far back as 19 April 1960. While the A-6E variant equipping squadrons in the 1980's was still a capable strike platform, having undergone numerous upgrades since the days of the A-6A, the basic design was getting somewhat 'long in the tooth' and was becoming increasingly vulnerable to modern surface to air threats. In addition, maintenance of the ageing airframes and engines was becoming more of a problem with each passing year.

At this time USAF research and development funding was paying dividends with major breakthroughs in 'stealth' technology for future aircraft designs. While ultimately the USN wanted to get into the 'stealth' aircraft business, its urgent requirement for a stopgap replacement for at least a part of its A-6E fleet required a more readily available solution.

The obvious stopgap answer to replace the A-6 appeared to be a new A-6, with new engines and modern avionics and weapons systems. While the USN pressed on with its plans for an advanced 'stealthy' strike aircraft under its ATA (Advanced Tactical Aircraft) program it was decided that the service should procure a number of new build A-6F Intruder II's, pending the availability of the ATA in the mid to late 1990s.

The A-6F prototype (BuNo.162183) made its first flight on 26 August 1987. However, the A-6F program was abruptly cancelled in 1988 when the Pentagon decided that the A-6E should continue in service until replaced by the ATA in the mid-1990's.

Following the cancellation of the A-6F, Grumman proposed a more modest Intruder update known as the A-6G, which would have incorporated most of the systems planned for the A-6F. In reality the A-6G was 'doomed' from the start as the USN was committed to procuring a 'stealthy' replacement for the Intruder through the ATA program and had no available funding for new Intruder variants.

Generic model of an early Lockheed concept for an ASTOVL strike fighter. Lockheed Martin

The ATA program was revealed in 1985 as a 'stealthy' subsonic all weather deep penetration strike aircraft, although the program had been underway for some time prior to this to meet a USN requirement for a stealthy replacement for the A-6 from the mid-1990s. As the USN had diverted little funds to the development of stealth technology the ATA drew on much of the technology developed with USAF funding. At the same time, the USAF was pursuing the Advanced Tactical Fighter (ATF) program (eventually won by the Lockheed Martin/Boeing F-22 Raptor). In April 1986, a MoU (Memorandum of Understanding) was signed between the USAF and the USN under which the former would consider the ATA as a potential replacement for its General Dynamics F-111 and later F-15E strike aircraft. It was projected at this time that the USAF could expect delivery of its first ATA in 1998.

Both services embarked on a joint program to develop a new engine that could be suitable for both the Navy's ATA and the air forces ATF. The engine development program, which was known as the JAFE (Joint Advanced Fighter Engine) program, eventually produced the Pratt & Whitney YF119 and rival General Electric YF120 engines for the USAF ATF competition. The USN eventually withdrew from the JAFE program when it became clear that the ATF engines would not be suitable for integration into the ATA.

Only two-teams, General Dynamics/McDonnell Douglas and Northrop teamed with Grumman and LTV responded to the USN request for proposals for its ATA. On 20 December 1987, the Northrop/Grumman/LTV team declined to submit 'best and final' bids for the ATA and four days later on the 24th, the General Dynamics/McDonnell Douglas team was declared the winner of the competition with General Dynamics as prime contractor.

The Advanced Tactical Aircraft, which was designated A-12 was designed as an extremely stealth driven flying wing, two-seat strike aircraft capable of carrying a large payload in an internal weapons bay, with a combat radius of 1850-km (1,000-nm) without in-flight refuelling. The A-12 was a more 'stealth' driven design than the USAF ATF competitors, the YF-22 and the YF-23, incorporating more RAM (Radar Absorbent Material) in its construction. Once the USN had withdrawn from the JAFE program it decided to power the A-12 with a pair of up-rated non-afterburning derivatives of the General Electric F404 turbofan, which powered the F/A-18A/B/C/D. This power plant was originally known as the F404/F5D2, but later matured into the F412-GE-400.

The A-12 was to be a subsonic aircraft with no real STOL (short take-off and landing) capability. The strike mission required that the aircraft be capable of penetrating enemy defences at ultra low level in all weather and in a dense ECCM (Electronic Counter Counter Measures) environment. Early in the program take-off weight was estimated at around 25000-kg (55,000-lb), although as the program developed, weight grew to the point where the contractors were experiencing difficulty in keeping within the specification weight limit of 60,000-lb.

The radar developed by Westinghouse was to have incorporated ISA (Inverse Synthetic Array) technology which was also planned for the Norden Systems APQ-173 radar developed for the A-6F Intruder II.

Had fortune shone more brightly on the troubled A-12 it was to have begun replacing the A-6E fleet from the mid-to-late 1990's, with at least 450 aircraft required for the USN, while up to a further 60 would have went to the USMC. If the A-12 proved to be too expensive for the USMC then the F/A-18D 'Plus' would be purchased as an A-6E replacement. The USMC eventually procured a 'night attack' two-seat variant of the F/A-18D for the strike role.

A model of a 1993 McDonnell Douglas Advanced Short Take-Off and Vertical Landing (ASTOVL) concept undergoing wind tunnel testing at the NASA Lewis Research Centre. NASA

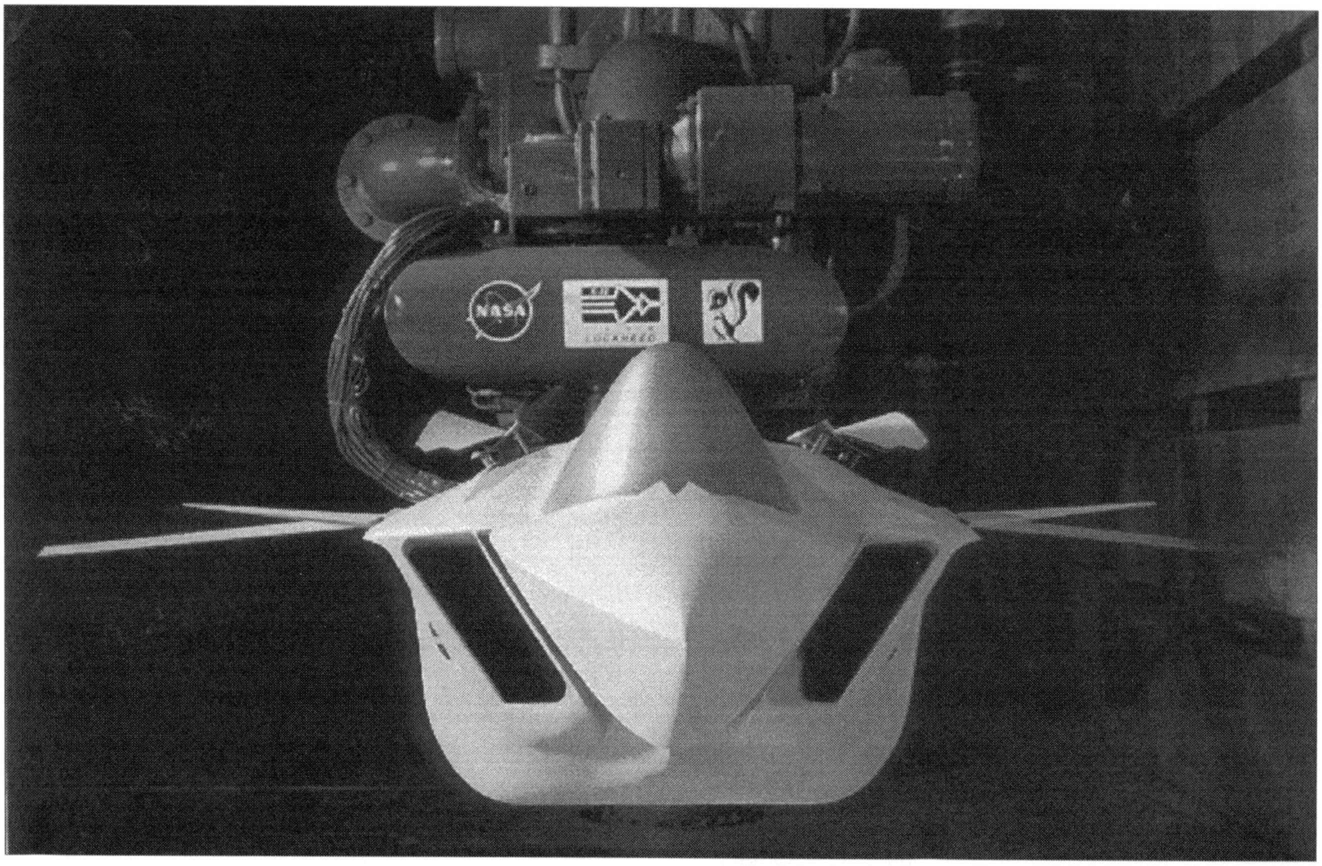

Above: This model of a 1990's Lockheed ASTOVL concept undergoing tunnel testing at a NASA Research Centre showed clear stealth design heritage in the forward fuselage and intake areas Like the Large Scale Powered Model (LSPM), which emerged in the mid-1990's this model featured forward canards. The canopy looks like it has been borrowed from the F-22 Raptor design. Lockheed Martin **Right: A-X jet effects model in the NASA Langley 16-foot transonic wind tunnel on 23.4.94. Like so many of its forebears, the A-X was cancelled.** NASA

The first A-12 prototype, BuNo.164519, was initially expected to make its first flight in 1990, but program delays saw this re-scheduled for mid-1991, and then again for March 1992. On 11 June 1990, US Secretary of Defence Dick Chaney had summoned the chairman of General Dynamics and MDC for an explanation as to why the A-12 delays had escaped the Major Aircraft Review (MAR) in spring 1991. USN sources claimed that the contractors were experiencing difficulty in building the A-12 to the required specification weights, which set out a loaded weight of 60,000-lb; slightly more than the A-6E which it was to replace.

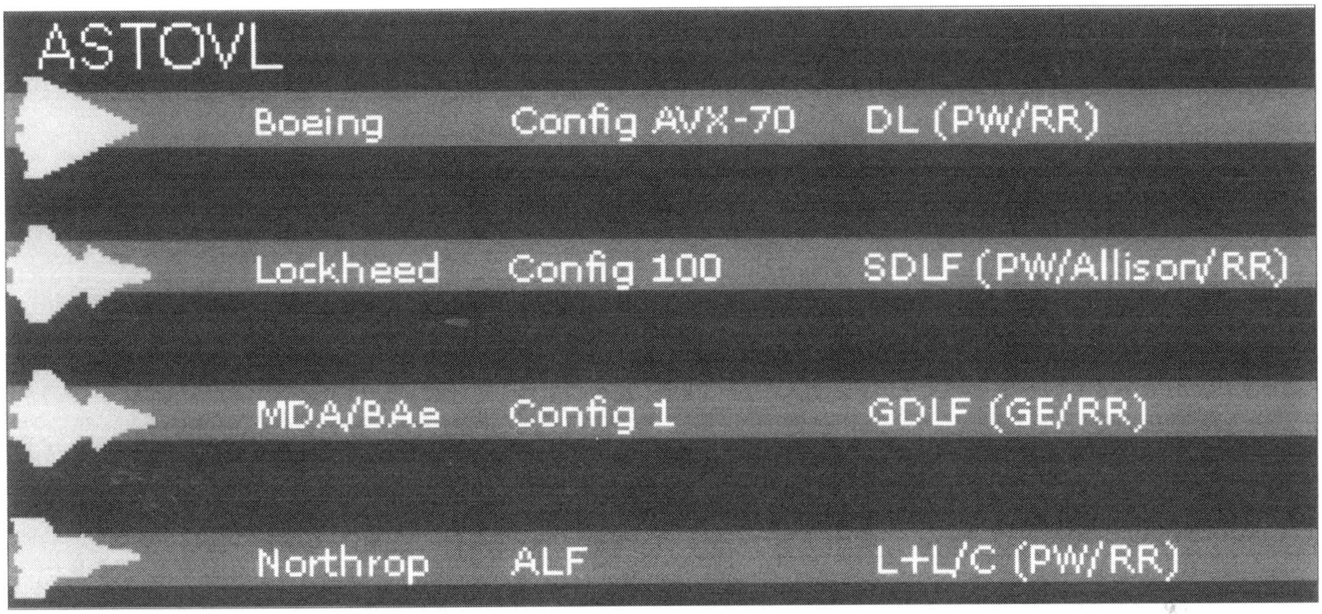

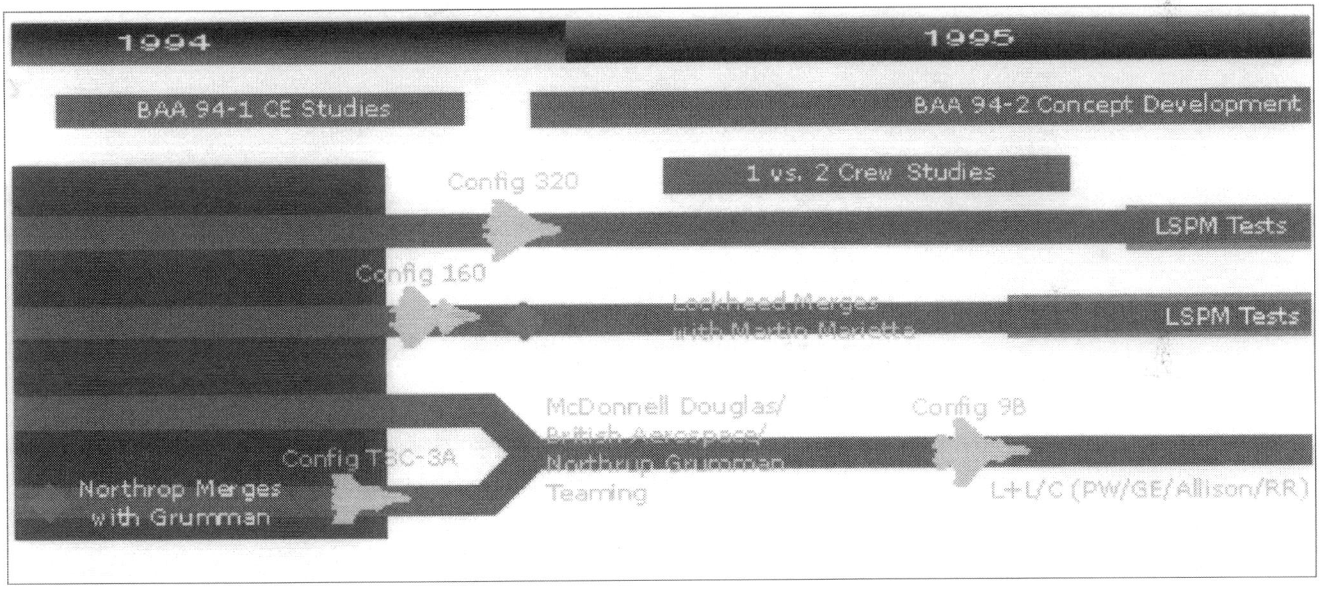

Top: Before the ASTOVL program was absorbed into the JAST program the four main teams had designs on the drawing board. Boeing was pushing the AVX-70, a pure delta design that would have incorporated a Pratt &Whitney/Rolls Royce direct lift propulsion system. The Lockheed Martin Config 100 was a canard delta, incorporating a Pratt & Whitney/Allison/Rolls Royce Shaft Driven Lift Fan STOVL propulsion system. The McDonnell Douglas/British Aerospace teams Config 1 design was also a canard delta design, incorporating a gas-coupled remote lift fan STOVL propulsion system. The Northrop ALF incorporated a Lift + Lift Cruise STOVL propulsion system. JSFPO Above: **The chart above shows the evolution of the various JAST designs during the BAA 94-1 Concept Evaluation studies and the BAA 94-2 Concept Development. Boeing and Lockheed Martin built large-scale powered-models of their respective Config 320 and Config 160 designs. Northrop Grumman dropped its Config TSC-3A and merged with the McDonnell Douglas/British Aerospace team. However, the MDC/BAE/Northrop Grumman team did not build a large-scale powered-model of their Config 9B lift+lift/cruise design.** JSFPO

In December 1990, funding for additional new build F-14A Tomcat fighters as well as funds for upgrades from F-14A to F-14D was blocked. USN supporters of the Tomcat had hoped to upgrade the aircraft further, giving it a strike role, which would at least partly compensate for the pending withdrawal of the A-6 fleet. However, Chaney announced that he was instead calling for the allocation of some $3.5 billion in funding to develop an upgraded F/A-18 Hornet, which would eventually emerge as the F/A-18E/F Super Hornet.

While Tomcat production was halted from the mid-1990's, the F-14 increasingly took on the strike role armed with precision strike weapons as partial replacements for the capability lost with the retirement of the A-6E in 1996.

The axe finally fell on the A-12, which was cancelled outright on 7 January 1991. Following the A-12 cancellation, most of the major aircraft companies were knocking on the Pentagons door in desperate attempts to sell variants of existing hardware to replace the A-12. Long before the demise of the A-12, these same companies had continued work on advanced developments of their existing fighters in case of either delay or cancellation of the ATA program. As far back as July 1987, Casper Weinberger, then US Secretary of Defence, told the USN and USAF to look at upgrades of their respective fleets of McDonnell Douglas F/A-18 Hornet, F-15 Eagle and General Dynamics F-16 Fighting Falcons. This was in case of delays, cuts or even cancellation of the ATA or ATF programs. To this end, McDonnell Douglas had been working on an advanced Hornet variant designated Hornet 2000 and had looked at a number of options including variants fitted with a dorsal fuel tank and a completely redesigned new variant featuring an arrow shaped wing with canard fore-planes. The Hornet 2000 program was axed in 1988, although MDC continued its studies on a private basis.

Development of an advanced strike aircraft continued under the A-X program. However, as this was unlikely to be available before 2005/6, by which time a large number of the USN's fleet of A-6, F-14 and early build F/A-18's would be beyond their life expectancy. It had become clear that an interim fighter would have to be procured as a stop-gap measure, with MDC proposing its study of an upgraded Hornet, which would be a stretched and improved variant of the Hornet 2000. In the spring of 1991, the first basic impressions of the new Hornet variant, already designated F/A-18E/F appeared in the press. Throughout 1991, the F/A-18E/F design was further developed and refined.

As an alternative to the F/A-18E/F Grumman proposed a number of new Tomcat variants beginning with the F-14 'Quickstrike', which was basically an upgraded variant of the F-14D optimised for the strike role. More radical changes were incorporated into the Super Tomcat 21, a more advanced update of the 'Quickstrike', as a cheap alternative to the expensive and unaffordable Naval Advanced Tactical Fighter (NATF). The Attack Super Tomcat 21 was an evolution of the Super Tomcat 21 dedicated to the strike role, although it would have retained an air to air capability. A further upgrade of the Tomcat that was proposed as an NATF alternative was the ASF-14, which would have had the power plant and systems of the ATF in a refined Tomcat airframe.

Top: As the USN requirement for a strike aircraft to replace the ageing A-6E Intruder II's dragged on, a number of contenders were put forward including the Tomcat 21 from Grumman. Grumman **Centre: Lockheed Martin pushed the AF-117X, which was a navalised variant of the F-117 Nighthawk.** Lockheed Martin **Above: The USN eventually procured the Boeing F/A-18E/F Super Hornet, which was a 25% larger variant of the Boeing F/A-18A/B/C/D.** USN

Lockheed Martin unveiled its 86% scale powered model of its 1994 JAST design at the Skunk Works, Palmdale, California on 27 April 1995. The model, which was made of glass reinforced plastic and steel was re-assembled at NASA Ames Research Centre during May 1995. NASA

Following a meeting of the defence acquisition board on 5 May 1992, the US DoD gave the go-ahead for the F/A-18E/F program and an EMD (engineering manufacturing development) contract valued at $4.88 billion was awarded on 1 June that year. The procurement program for 1,000 aircraft was at that time estimated at $49 billion and was to last until 2014/15. However, the program was cut to under-500 aircraft.

Even in 1992, the program was under severe criticism as providing only marginal improvements over the existing Hornet at a large cost. Some believed the F/A-18E/F was depriving the USN of funding for the true A-12 replacement, the A-X, which was being studied as a low-observable complement to the F/A-18E/F and as such was seen by many as the true A-6 replacement. In parallel, the USAF had embarked on the MRF (Multi-Role Fighter) program in search of a replacement for its large fleet of F-16 and A-10 Thunderbolt II tactical fighters and ground attack aircraft. In mid-1992, the HASC (House Armed Services Committee) FY93 budget proposal proposed acceleration of the USN A-X program, requiring that existing sub-systems such as avionics and engines developed for other aircraft be used, and that two competitive prototypes be built by rival teams before selection of a winning design. This was to be at the expense of the F/A-18E/F program, resulting in the USN request for $1.1b in funding being cut by $535m.

The A-X study later broadened into the A/F-X, as a multi-role strike fighter replacement for not only the USN's A-6, but also the USAF's General Dynamics F-111E/F, McDonnell Douglas F-15E Eagle and Lockheed F-117A Nighthawk strike aircraft, with all the usual suspects involved. The General Dynamics/McDonnell Douglas team had planned to bid a variant of the cancelled A-12 Avenger II to meet the A/F-X requirement. However, following Lockheed's acquisition of General Dynamics in 1992/3, the company disbanded the General Dynamics design team working on the A/F-X as Lockheed was designing competing AF/X designs, including a variant of the winner of the USAF ATF contest, the Lockheed/Boeing F-22. Lockheed also proposed a variant of the F-16 designated F-16X to meet the USAF MRF requirement.

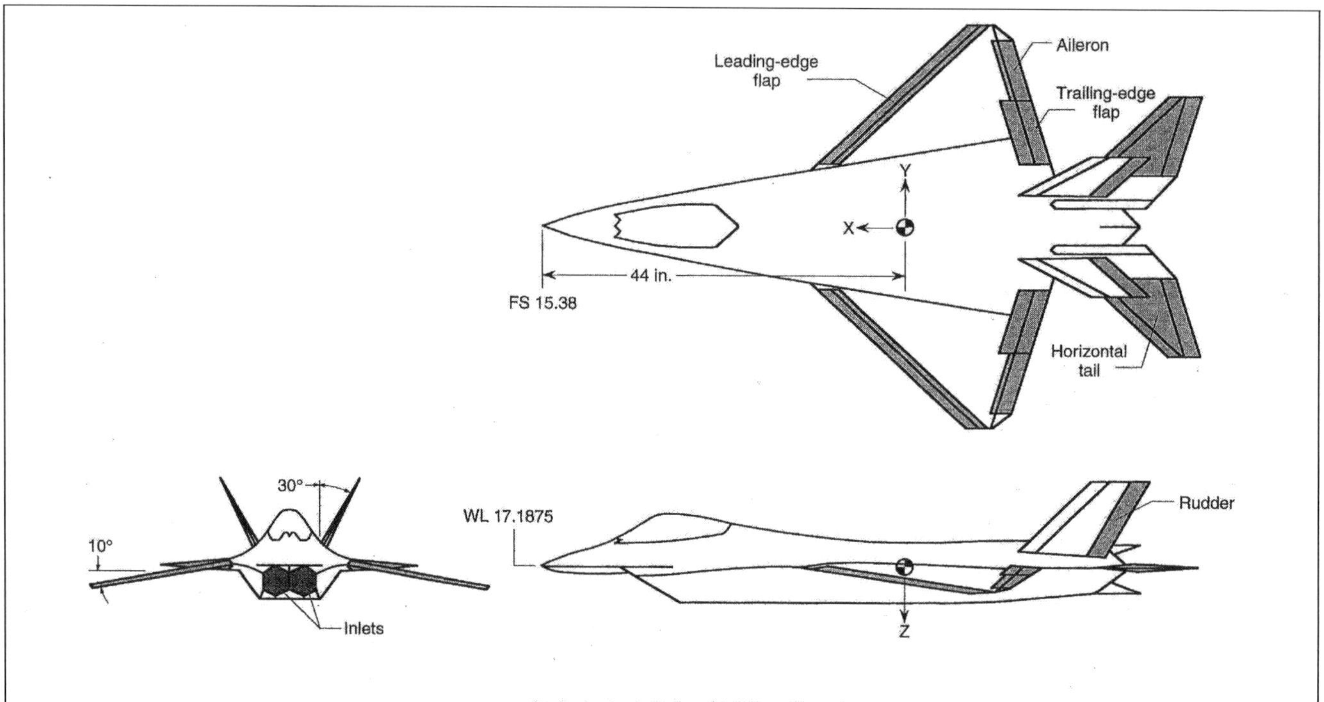

Boeing Multirole Fighter (BMRF) configuration.

Above and below right: Early 1990's Boeing concept designs for the MRF program, which was cancelled in 1993. Although clearly a much sleeker design than that adopted for the later X-32 JSF Concept Demonstrator, the Boeing Model 24, along with other stillborn designs like the tailless MRF-24X served as the baseline from which the X-32 evolved. NASA

Lockheed claimed that the F-16X, which would have featured a new wing, based on that of the F-22 and with no horizontal tail, would have twice the range of the F/A-18E/F at only two thirds of the cost.

In August 1993, the Pentagon's Bottom-Up Review (BUR) of defence recommended that the USN/USAF A/F-X and the USAF's MRF be cancelled, while at the same time confirming early retirement plans for the A-6E by the year 2000. (The A-6E was eventually retired in 1996.) In addition the BUR called for curtailment of F-16C/D and F/A-18C/D procurement and initiation of the JAST program. Later in the 1990's additional F-16's were procured. On the positive side, it recommended continuation of the F/A-18E/F for the USN and the Lockheed/Boeing F-22 for the USAF. The review also nominated the F/A-18E/F as a replacement for the USAF Lockheed F-16 and Fairchild A/OA-10A, and also set the future carrier fleet at 11 active vessels. This was later reinstated to twelve vessels, one of which would be in reserve, with ten-active and one reserve air wings. In addition to continuation of the F/A-18E/F and F-22 programs, the review recommended that both services combined with the USMC began planning for a future common strike fighter with an in service date of around 2012. This latter program would eventually result in the JAST (Joint Advanced Strike Technology) program.

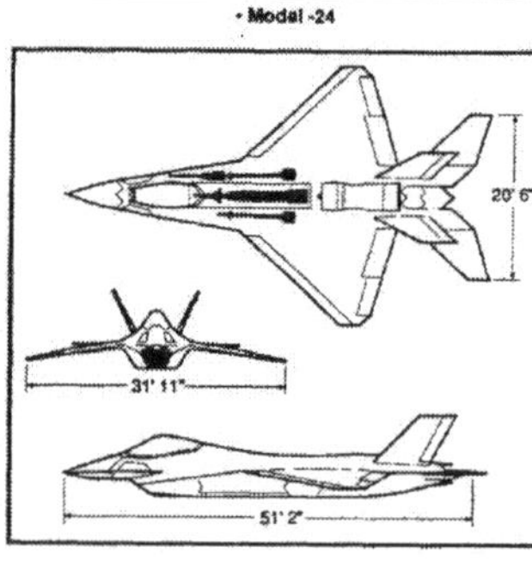

General
- 1998 technology, 2005 IOC
- Single crew
- FDWT: mission TOGW – 0.5 internal fuel
- Design LF: 9g @ FDWT, GR/TP structure
- Q-placard: 2,130 psf, M1.2 @ S.L.
- Maximum internal fuel capacity (lb) 8,690
- Installed avionics (lb) 1,598

Weights
- Takeoff gross weight 34,720
- FDWT 25,460
- Operating weight empty 19,980
- Mach, combat/max 0.9 / 2.2

Propulsion (lb)
- Sea level static A/B, installed (lb)
- T/W @ takeoff gross weight
- Nozzle 2-D/C-D, TV
- Inlet Fixed

Geometry
- Wing area, ref (sq ft) 465
- W/S @ takeoff gross weight (psf) 74.7
- Wing aspect ratio/taper 2.20 / 0.13
- Wing sweep, LE/TE (deg) 47.5 / 17.0
- Wing t/c @ (SOB/tip) (%) 4.5 / 3.0

Baseline Aircraft

In 1993, DARPA initiated a program aimed at development of a supersonic STOVL fighter to replace the AV-8B. It was around this time that the US DoD began to seriously consider cancellation of the A/F-X. A Joint Attack Fighter (JAF) was being proposed as a replacement for the A/F-X program. This aircraft would not only be cheaper than the A/F-X, but would be designed with a common airframe, suitable for all three fast jet users of the US armed services. One of the main attractions of such a program was the massive cost saving that could occur. Much of the philosophy surrounding the JAF, such as the single-engine single-crew and unprecedented level of commonality between variants, was later carried over to the JAST program.

Once the A/F-X had been cancelled most of the USN opponents to the F/A-18E/F came into line, thankful now for whatever they could get, while at the same time the USAF vehemently rejected any notion of the F/A-18E/F being adopted as an F-16 replacement. With the A/F-X now consigned to oblivion like the A-12 before it, Lockheed revived its proposal for a carrier-based variant of the F-117 as an A-6E replacement. The original proposal designated F-117N (N for naval) would have retained the same basic design of the USAF F-117A, although stronger materials would have been added to allow the aircraft to be operated in the demanding carrier operational environment, as well as an off-the-shelf automatic carrier landing system (ACLS). Following its 1992 decision to develop the F/A-18E/F as its future strike fighter, the USN told Lockheed to stop pushing the unwanted AF-117. Lockheed, however, pretty much chose to ignore this and continued with development of a variant of the F-117 more suited to operations from a carrier deck.

Lockheed redesigned the F-117 for the naval role and began proposing a new variant designated A/F-117X. This new variant would have a similar maximum take-off weight to the proposed F-117Bs maximum take-off weight of 33203-kg (73,200-lb). It would also have had a larger payload and be powered by the General Electric F414 powerplant being developed for the F/A-18E/F. This powerplant was also proposed for the F-117B being offered to the USAF. The A/F-117X would have been fitted with the F-14 Tomcats undercarriage and the fuselage would have been strengthened to take the stress of carrier operations. The most obvious external differences would be the new trapezoidal horizontal tail, reworked exhaust area and a new bubble canopy based on the canopy designed for the Lockheed/Boeing F-22. An arrestor hook would have been fitted for carrier operations and the aircraft would have had folding wings. Lockheed claimed an AIM-120 AMRAAM and AIM-9 Sidewinder capability for the A/F-117X, which was to be equipped with unspecified radar. While the A/F-117X would undoubtedly have been a competent strike platform it had little support in the Pentagon and even less support in the USN. Therefore, it was ultimately doomed; joining the A-6F, A-6G, A-12, A-X, A/F-X and various advanced Tomcat variants in the league of USN could have beans.

On 27 January 1994, the JAST program office was formally established with the goal of defining and developing aircraft, weapons and sensor technology that would support the development of future tactical aircraft. Eventually the program moved from "a broad, all-encompassing program to one that would develop a common family of aircraft to replace several US and UK aircraft." By the end of 1994, the CALF (Common Affordable Lightweight Fighter) was absorbed into the JAST program. CALF was renamed ALF, then became the primary focus of Joint Advanced Strike Technology (JAST) program. As JAST was looking at STOVL, CTOL and CV aircraft, Congress mandated that JAST and CALF be combined into a single program. As both programs appeared to be studying similar concepts with similar goals this merger was accomplished with little disruption.

Top: This drawing of the proposed MDC/BAe Config 1 shows the canard-delta layout, featuring a gas-coupled remote lift fan STOVL propulsion system. Above: During the Concept evaluation phase, Northrop became Northrop Grumman, dropping its Lift + Lift Cruise STOVL propulsion system Config TSC-3A, which had evolved from its ALF design. It then merged its program with the MDC/BAe team, which now put forward the Config 9B lift+lift/cruise design, which had lost the canard fore-planes and had no conventional vertical tail-planes. Instead, the horizontal stabilators were upturned at an angle of 25°. Split ailerons on the wing trailing edge were used as rudders and air brakes. Both JSFPO

Under a six-month Concept Exploration (CE) study awarded in May 1994, the JAST program initially explored a wide-range of potential strike aircraft concepts, the findings of which showed that a tri-service family of aircraft was considered the most affordable solution to meet the needs of the three US services. This aircraft family would be based on a single common airframe design with three distinct variants. The CTOL variants would be developed for the USAF to replace

Boeings single 96% scale Large Scale Powered Model (LSPM was powered by a single YF119 turbofan engine. This model was used for hover and engine tests, which focused on near-ground effects, including: hot gas ingestion; ground effects; suck-down and thermal and acoustic footprints. Boeing

the F-16 and A-10; the STOVL for the USMC would replace the AV-8B and F/A-18 Hornet and CV for the USN would complement the F/A-18E/F.

After numerous trade studies, two critical decisions were made. The aircraft would be a single-engine, single-crew strike fighter. Traditionally, the preferred choice for navy fighters has been twin-engine designs, in case one fails over the sea. However, it should be noted that the single engine choice for an USN fighter is hardly unique, as many recent and current fighter and attack aircraft within the USN and other navies have been powered by a single-engine. Acceptance of a single-crew aircraft was subject to continued studies and appropriate technology maturation.

In December 1994, Lockheed Martin, McDonnell Douglas, Northrop Grumman and Boeing were each awarded fifteen-month Concept Definition and Design Research (CDDR) contracts. Northrop Grumman, McDonnell Douglas and British Aerospace teamed after award of the CDDR contracts. Following award of the CDDR the various contractors refined their Preferred Weapon System Concept (PWSC) designs and carried out a number of risk reduction studies such as wind-tunnel tests, powered model STOVL tests and engineering analysis.

In the spring of 1995, all contractors selected the Pratt & Whitney F119 engine to power their respective aircraft designs. This resulted in Pratt & Whitney being awarded a contract in November 1995, for the preliminary design of each JAST engine concept. General Electric was awarded a contract to investigate whether the F110 or the F120 would be suitable for development as an alternative engine for one or more of the three aircraft variants. In 1996, the conclusion of these studies revealed that the F120 had been identified as the most suitable choice for an alternative engine and General Electric initiated preliminary design efforts.

On 9 August 1994, a MoU had been signed by the US ARPA (Advanced Research Projects Agency) and the United Kingdom MoD, which covered UK participation in the CALF program. UK participation was to be in Phase II; the research and technology demonstration phase of the program, which had commenced in March 1993. The UK's financial contribution to Phase two was $12 million, to provide technology demonstrations of critical high-risk designs as well as research into integrated flight and propulsion systems. The USAF allocated the proposed demonstrator aircraft the X-32 designation. Three teams were bidding for two concept-demonstration contracts. The Teams were Lockheed

Martin, Boeing and McDonnell Douglas/British Aerospace/Northrop Grumman.

A draft Request for Proposals (RFP) was issued in December 1995, followed in February 1996 by a final RFP. On 20 December 1995, the UK and USA signed a MoU committing the UK to take part in the 4-year concept demonstration phase of the JAST program. Under this MoU, the UK would contribute £130m ($200m); some 10% of the concept demonstration phase cost to the project.

The Boeing approach to meeting the STOVL requirement was to use a variant of the tried and tested direct lift STOVL system used in the BAE Systems Harrier, Sea Harrier and the Boeing/BAE Systems Harrier II subsonic STOVL combat aircraft. Boeing

McDonnell Douglas and BAe were originally looking at powering their JSF technology demonstrator contender with the GE-YF120 engine, which was unsuccessful in the US Advanced Tactical Fighter (ATF) competition, which eventually saw the Lockheed Martin Boeing F-22 Raptor powered by the F119 engine winning over the Northrop/McDonnell Douglas YF-23. MDC and BAe later abandoned this concept and reverted to the P&W F119 lift plus lift cruise powerplant. Lockheed Martin and Boeing also chose the F119 for their concept demonstrator proposals, although General Electric was to have the opportunity to compete to power production aircraft and was studying derivatives of both the F110 and the YF120. During September 1995, it was revealed that GE was to be allowed to compete with P&W on the powerplant choice of the CTOL variant of JSF, with a later decision that it could compete to power all three variants of the JSF, CTOL, CV and STOVL, with the exception of the first production batches. The STOVL propulsion system developed by Rolls Royce would be common to both engine types being developed for the JSF.

General Electric, with Allison engines was offering JSF airframe designer's growth variants of the F110 engine, which powers variants of the F-14 and F-16, and the YF120. Two F110 growth steps were offered. The first was the DF3/DF6 providing 20-25% more power than the 29,000-lb (130-kN) F110-129 by using the uprated core from the CFM56-7 commercial turbofan. The second was the DV2/DV3 providing 35-40% more thrust through the use of a higher-airflow bladed-disc (blisk) fan and variable cycle-core. Two variants of the YF120 were proposed: the V4 providing 20-25% more thrust, with a blisk fan and variable-cycle core; and the F3 providing 40% more thrust, with a fixed cycle core. In May 1996, it was announced that the YF120 had been chosen as the alternative power plant for the JSF, although plans would wax and wane in the years to come.

Originally, large-scale powered models (LSPM) from all three JAST contenders (Boeing, McDonnell Douglas/Northrop Grumman/BAe, and Lockheed Martin) were scheduled for testing at Ames facilities during the initial stage. However, only two of the competing teams, Lockheed Martin and Boeing built the large-scale models. The Lockheed Martin model was involved in the tests in the Ames Outdoor Aerodynamic Research Facility (OARF) and the 80-ft by 120-ft wind tunnel test section. The powered aerodynamic testing on the model was successfully completed on 28 February 1996. As well as providing its exceptional facilities to the program, Ames also contributed research and operations engineering expertise.

Lockheed had unveiled its 86% scale powered model of its 1994 JSF design at the Skunk Works, Palmdale, California on 27 April 1995. The model was constructed primarily of steel, with fibreglass being used in low-stress areas. It had movable control surfaces, including canard fore-planes, flaps and rudders that were pre-positioned for each run. To provide vertical flight capability the Lockheed Martin concept incorporated a forward fuselage-mounted vertical lift fan in addition to an aft-mounted vectoring nozzle, all driven by the single F100 engine. Power was delivered from the engine to the lift fan by means of a shaft and gear system.

The 86% scale was dictated by the need to match the thrust of the Pratt & Whitney F100-P&W-229 plus engine simulating the larger F119 powerplant planned for the full size JSF demonstrator. The vertical lift fan, which was located aft of the cockpit on the spine of the aircraft was the distinctive design feature. Immediately aft of the lift fan was a large rectangular auxiliary inlet for the F100-229 engine. Air from the compressor was

Boeings 96% scale YF119 powered JSF Large Scale Powered Model (LSPM) was assembled at the Defence and Space Groups Kent site in Washington. It was then moved to the new purpose built site at nearby Tulalip for hover and engine tests, which commenced in September 1995 and concluded in December that year, at a JSF-dedicated facility. Boeing

fed from a Rolls Royce duct system to simple roll-control nozzles buried in the root of each wing. The main engine exhaust was a Rolls Royce built two-dimensional lift/cruise nozzle with thrust vectoring and variable-area control. The nozzle could vector up to 10-degrees up for increased manoeuvrability and up to 20-degrees forward to aid deceleration for an approach to vertical landing.

The model was re-assembled at NASA Ames Research Centre during May 1995, with attachment of the 9.1-m span wing, canard fore-planes and twin canted vertical tails prior to hover tests. The Ames contractor team conducted two test series of the LSPM, one at the OARF and the other at the 80-ft by 120-ft wind tunnel. Goals of the OARF test were to obtain engine thrust calibrations and aerodynamic forces during hover, both in and out of ground effect. Some of the hover data was later used for comparisons between small and large-scale model testing.

The three-month OARF test program began on 10 July 1995 was successfully concluded on 27 October that year. For the tests, which evaluated jet thrust, airframe, suck-down, hot gas ingestion and ground pressures and temperatures, the model was suspended from the ground by grabs attached to large hooks on the aircraft's upper fuselage, with the tests reportedly producing better than expected results.

After the OARF testing the model was tested in the 80-ft by 120-ft wind tunnel at Ames. The main goal of this testing was to obtain data for evaluating differences between small and large-scale model testing during the transition flight regime between jet borne and wing-borne flight. Aerodynamic testing of the model in the un-powered configuration occurred from 27 November through 10 December 1995. Powered aerodynamic testing of the model began on 11 December, and was successfully completed on 28 February 1996. In all, over 120 hours of engine-on and 60 hours of engine-off testing were completed.

Since the STOVL version of the JAST aircraft would be required to hover, the propulsion system had to provide thrust in the vertical direction to support the

Above and right: Although it was unable to disclose facts at the time, it emerged in early 2003 that Boeing had used data gathered during testing of the stealthy Bird of Prey demonstrator aircraft in its Joint Strike Fighter bid. Although to what extend this influenced the design of the X-32 CDA and the later planned Preferred Weapon System Concept is unclear. The Bird of Prey was flight tested during 1996. Both Boeing

aircraft weight. This thrust vectoring caused complex jet flow fields around the aircraft that were critical to its performance and flight worthiness. Current (then) mathematical predictions and small-scale testing could not adequately predict these interactions, but a large to full-scale model could. This size of model more closely resembled a flight-worthy vehicle, without the high costs associated with flight-testing. Ames' facilities, encompassing the 40-ft by 80-ft and the 80-ft by 120-ft wind tunnels, provided the only place that transition (to and from wing-borne flight) testing of these large- to full-scale models could be adequately accomplished.

Lockheed Martin also experimented with other techniques for reducing rollout in a short landing, including using the canard foreplanes as air brakes. It was discovered that aerodynamics was not very effective at STOVL speeds. The slight anhedral of the delta wing was offset by an equivalent dihedral angle of the fore-planes, ensuring that the aircraft's fore-planes could not interfere with the airflow over the wing. The canard fore-planes were eventually dropped in favour of a more conventional tail-plane seen on the X-35.

Lockheed Martin turned to Yakovlev in Russia to discuss ASTOVL technology transfers in mid 1995 after being refused access to BAe ASTOVL technology due to BAe's commercial agreement with MDC. Yakovlev had built and flown a number of STOVL aircraft including the Yak-38 ‘Forger’ which had entered operational service, operating in a similar fashion to

In late 1995, the McDonnell Douglas/Northrop Grumman/British Aerospace team unveiled its revised JAST contender, Configuration 9B. The team had switched from its original concept of a gas-coupled remote fan STOVL fighter to a lift-plus-lift/cruise STOVL concept. Gone completely was the forward canard of Configuration 1 from 1993, which was replaced with a tailed design with a lambda-shape main wing configuration. The team opted to forego conventional vertical tail-planes with the horizontal stabilators being upturned at an angle of 25-degrees. MDC (Boeing)

British Sea Harriers and the supersonic Yak-141. However, when BAe joined the Lockheed Martin team in 1997 it brought with it a wealth of experience in STOVL operations and design, therefore, negating the need for technology from the Russian manufacturer.

Boeings 96% scale YF119 powered JSF Large Scale Powered Model (LSPM) was assembled at the Defence and Space Groups Kent site in Washington, before being moved to the new purpose built site at nearby Tulalip for hover and engine tests, which commenced in September 1995, and concluded in December that year. These tests focused on near-ground effects, including: hot gas ingestion; ground effects; suck-down and thermal and acoustic footprints as well as some 5,300 hours of wind-tunnel and other testing before mid-1996.

Boeing adopted a blended Delta configuration for its initial JSF design, on the basis that the blended Delta produces a large internal fuel volume. This would enable a range/payload at least double that of any of the aircraft to be replaced by JSF. Fuel fraction was 40%. More fuel would be carried internally than any of the current fighter/attack aircraft equipped with external fuel tanks, allowing a range/payload about 2.5 times that of the F-16/F/A-18 class of combat aircraft.

Recognising the need to validate its innovative JSF concept, concurrent with engineering and program milestones, Boeing conducted an unprecedented series of simulations and tests of its JSF design. Numerous component tests were followed by testing of a full-scale model of the then proposed operational aircraft in the Boeing Indoor Radar Range at the Seattle Developmental Centre. These tests, which were completed in April 1996, were so successful that Boeing needed only a few hours to validate the design instead of the weeks originally scheduled to match results with detailed pre-test predictions. The radar range tests strongly validated not only Boeing's unique design, but its basic analysis and test methodology as well. Using the large-scale powered model mentioned above and small-scale models, propulsion tests verified propulsion component performance and propulsion elements in key

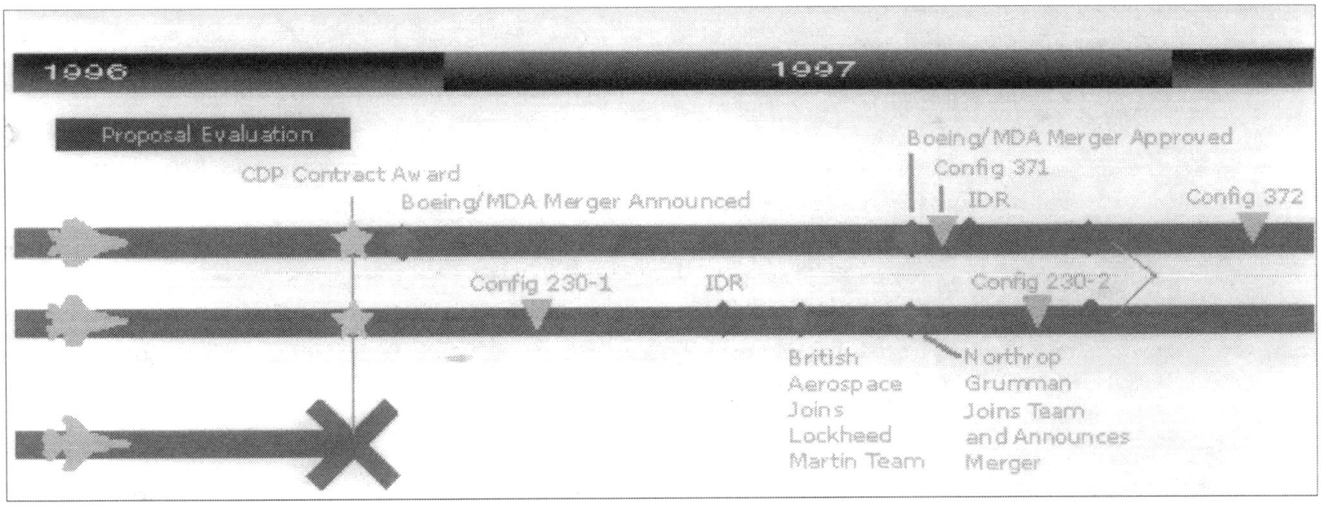

The chart above shows the developments and mergers from the CDP contract awards in late 1996 and into 1997. The McDonnell Douglas Aircraft/British Aerospace/Northrop Grumman team was eliminated from the program when the CDP contracts were awarded to the rival Lockheed Martin and Boeing teams. MDA was then absorbed into Boeing in a merger and BAe and Northrop Grumman joined the Lockheed Martin team. JSFPO

areas such as inlet, lift systems and exhaust systems. Low Speed Aerodynamics tests involved wind-tunnel tests to develop configurations and to establish the stability and control data to determine the configurations flying qualities. The tests were conducted in three separate low-speed wind tunnels. High Speed Aerodynamic tests focused on transonic and supersonic flight; these tests further refining the configuration and verifying vehicle performance. The tests were performed in six different wind-tunnel facilities in the United States and the United Kingdom. High Angle of Attack tests was performed at low speeds up to 90 degrees angle of attack and up to 50 degrees angle of attack at high speeds. The tests were conducted using sub-scale models in four different wind tunnels. An extensive ground effects tests series (including those with the LSPM) involved a combination of small-scale and large-scale models to determine hover, transition effects and hot gas ingestion. Weapons Integration tests involved aircraft and weapon models to measure weapon bay acoustic characteristics and confirm internal weapon bay separation predictions.

Among the more than 11,700 hours accumulated in the test and simulation were successful tests in facilities on two continents. Some of the major areas of the test program included.

- low speed aerodynamic and propulsion tests.
- high speed aerodynamic and propulsion testing.
- high angle of attack testing
- Short take off and vertical landing ground effects testing.
- handling qualities dome simulation.
- virtual; avionics prototypes dome simulation
- weapons integration and separation testing.

In July 1996, with more than 11,700 hours of successful tests to its JSF design, The Boeing Company submitted its proposal for the JSF. The Boeing emphasis was on simplicity in its JSF design to reduce the cost of building the aircraft and cut the customer's operations and supports costs. The simplicity of the Boeing JSF design was underscored by its having only four basic structural components: a single piece wing; integrated fore-body; centre fuselage and aft fuselage/tail. Instead of fastening two wings to the sides of a fuselage like most fighter aircraft, the Boeing concept called for the fuselage to be attached underneath a single composite wing. This would be the primary load bearing component as well as a very large fuel tank, giving the Boeing JSF an internal fuel capacity in excess of any of the aircraft it was designed to replace. Boeing also made extensive use of advanced technologies and processes, rapid prototyping, innovative design characteristics and integrated product teams.

In late 1995, the McDonnell Douglas/Northrop Grumman/British Aerospace team revealed details of its revised JAST contender, Configuration 9B, after switching from its original concept of a gas-coupled remote fan STOVL fighter to a lift-plus-lift/cruise STOVL concept. MDC stated that the change came about after the ASTOVL program was merged with JAST and affordability became a key requirement. Gone completely was the forward canard of Configuration 1 from 1993, which was replaced with a tailed design with a lambda-shape main wing configuration. As with the YF-23, the McDonnell Douglas/Northrop Grumman/British Aerospace team opted to forego conventional vertical tail-planes on its JSF contender. Instead, the horizontal stabilators were upturned at an angle of 25-degrees. This differed considerably from the 40-degree elevation of the horizontal stabilators on the YF-23. Split ailerons on the wing trailing edge were used as rudders and air brakes. The Configuration 1B design would have been powered by a relatively unmodified P&W F119 engine for cruise. The STOVL variant would have incorporated a blocker/diverter valve to deflect main engine thrust to a vertical lift nozzle under the rear fuselage, which along with the separate lift-engine would allow STOVL operations. Pitch and yaw control would have been provided by thrust vectoring, however, the team claimed that the aircraft would have been controllable throughout the entire flight envelope on aerodynamic controls alone.

McDonnell Douglas/Northrop Grumman/British Aerospace considered lift engine bids from General Electric/Allison and Rolls Royce. The GE/Allison bid was based on a joint integrated high-performance turbine engine technology developed under its military engine teaming effort announced in July 1994, before the purchase of Allison by Rolls Royce. The Rolls Royce bid was based on lift engine work already carried out for Northrop Grumman under a $2m study contract. Rolls Royce initially studied mating the core of the EJ200 engine developed for the Eurofighter Typhoon, with the fan of a Rolls Royce Tay turbofan, but found that the core was too long to fit in the small fuselage area. The 16,000-lb (70kN) thrust lift engine was required to fit into an area measuring only 1.15 x 1.15m. Attentions are then thought to have turned to the de-staged BMW Rolls Royce BR700 core with a reduced pressure ratio. The engine would also have been offered with "squat" fan slim line guide vanes and a narrower combustion chamber. Based on earlier statements from Northrop - which until its bid merger with McDonnell Douglas was the only company with a lift plus-cruise JSF solution - the requirement called for a simple engine with a short duty cycle.

The team said that the advantage of the modular lift engine, which could be replaced with a tank holding 500-kg of fuel in the conventional take off versions, outweighed the disadvantages. These included the cost of the lift engine development, plus the reliability and maintenance impact of two engines. Advantages included more dedicated tailoring of engine requirements to meet lift and cruise, better aerodynamic integration, manufacturing modularity, and as demonstrated by Yakovlev, it was a proven short take off and vertical landing system. Additional fuel or a second cockpit in the CTOL variant could replace the lift engine in STOVL variants. By December 1995, MDC had completed almost half of its 6,500 hour aerodynamic and propulsion wind tunnel testing program. Originally the MDC design featured canard fore-planes with a wing of trapezoidal design but this was later changed to a Lambda shaped wing and conventional tail surfaces. The CTOL variant for the USN had increased span outboard of the wing fold and high lift devices for carrier suitability.

This model of the Boeing X-32B seen at Farnborough in September 96, was typical of the design that went into the concept demonstrator program. Author

At a press conference in Washington DC on 12 February 1996, the McDonnell Douglas led team unveiled its entry in the JSF competition. The aircraft presented was a sleek, near tailless design. The vertical tails having been deleted from the earlier design and an engine nozzle was added giving the aircraft thrust vectoring capability. The MDC team claimed that the resultant design was "a highly manoeuvrable aircraft that is virtually spin proof." Further stating that the teams JSF would achieve pitch (up and down), and yaw (left and right), control from a special nozzle mounted on the cruise engine that would deflect engine thrust.

The MDC/Northrop Grumman/BAe team's revised lift-engine exhaust design would eliminate a hot gas ground environment issue uncovered during rig tests of its JSF contender. The revised design would deflect exhaust for the lift at idle. Exhaust for the GEA-FXL lift engine to be developed by General Electric, Allison Engines and Rolls Royce, would exit through louvered vanes. In the hover, the louvers split to direct gas fore and aft. For the transition to forward flight the vanes vector thrust 45 degree aft and a 5 degree forward vectored braking capability was tested. The team remained confident that the 17/18,000-lb (75-80-kN) thrust lift engine was powerful enough to propel the aircraft in the event of a main cruise/lift engine failure. The wind-tunnel data appeared to support this.

Despite the lift engines proximity to the main engine inlets, MDC stated that hot gas ingestion was not a threat and that LIDS would not be required. LIDS running along either side of the belly on the original design to offset anticipated suck-down effects had been eliminated reducing weight and complexity. The team said that the thousands of hours of wind-tunnel tests confirmed that the designs flying qualities were

This artist impression depicts a late design McDonnell Douglas/Northrop Grumman/British Aerospace JAST concept from the CDDR. MDC

weapons would allow the aircraft to stay out of detection and or engagement range of enemy air defences; at least in theory. Of course, this theory was to a large extent based on outdated, obsolete air defence systems of the types faced in the 1990's.

On 20 December 1995, the UK and USA signed a MoU committing the UK to take part in the Phase III, four-year concept demonstration phase of the JAST program. Under this MoU, the UK was to contribute £130m ($200m); some 10% of the concept demonstration phase cost to the project. The JSF Program Office (formerly the Joint Advanced Strike Technology Program) based in Washington, DC, in early 1996, issued its Request for Proposals (RFP) for the concept demonstration phase of the program. In July 1996, the competing teams submitted their respective proposal for the JSF and in November that year the US DoD awarded Boeing and Lockheed Martin 51-month contracts to proceed with the JSF CDP (Concept Demonstration Program) leaving the third team of McDonnell Douglas/British Aerospace/Northrop Grumman as the losing contender. The contract awards, which were valued at approximately $660 million, mandated that the competing contractors build aircraft that would demonstrate three specific requirements: (1) commonality; (2) low-speed handling qualities and carrier approach and (3) STOVL hover and transition between conventional and vertical flight.

Ultimately, the contractors respective proposed weapon system designs, technology maturation work and the flight demonstration results were all to be considered in selecting the winning contractor in 2001, for the Engineering and Manufacturing Development (EMD) phase of the program and for production.

excellent and drag prediction tests confirmed analytical predictions.

While the McDonnell Douglas/British Aerospace/Northrop Grumman JSF design offered obvious advantages in terms of weight savings, reduced drag, and increased stealth effectiveness, it was always going to be at a disadvantage because it required two separate types of engine. This would add a considerable burden on maintenance and staffing levels, and none of the prospective customers needed to have an additional engine type in their inventories.

The Initial Joint Interim Requirements Document 1 (JIRD 1), which was released in 1995, set out the various parameters of the JSF. These included required speed, size, range and stealth, therefore, enabling the competing contractors to set the outer lines of their respective JSF candidates, which for Boeing was set at the model 370 delta wing and Lockheed Martin the configuration 230-1, which eventually became the X-32 and X-35 respectively. Three critical design areas set out for the manufacturers were lethality, survivability and capability. Exotic maximum speeds such as the Mach 2.5 attained by legacy F-15 generation aircraft was not part of the JSF requirement as the aircraft's stealth, qualities, advanced defensive aids and stand-off

Chapter Two

The Boeing Joint Strike Fighter

The structurally complete X-32A CTOL CDA is seen at Palmdale, California in late 1999. Boeing

From the outset, Boeing focused its JSF design concept on the need for the aircraft to be affordable and to meet multi-service requirements. The Boeing emphasis was on simplicity in its JSF design to reduce the cost of building the aircraft, and cut the customer's operations and supports costs. The simplicity of the Boeing JSF design was 'underscored' by its having only four basic structural components: a single piece wing, integrated fore-body centre fuselage and aft fuselage/tail. Instead of fastening two wings to the sides of a fuselage like most fighter aircraft, the Boeing concept called for the fuselage to be attached underneath a single composite wing, which is the primary load-bearing component. In addition, the wing is a very large fuel tank, giving the Boeing JSF an internal fuel capacity well in excess of any of the aircraft it was designed to replace.

Another of the major successful Boeing JSF design validations was completed in April 1996, at the indoor Radar Range in Seattle where Boeing engineers used a full-scale model of the company's JSF in RCS (radar cross section) tests. After conducting numerous component tests, Boeing began testing the full-scale model of PWSC in the Boeing Indoor Radar Range at the Seattle Developmental Centre. These tests, which were completed in April 1996, were so successful that Boeing needed only a few hours to validate the design, instead of the weeks originally scheduled to match results with detailed pre-test predictions. The radar range tests strongly validated not only Boeing's unique design, but also its basic analysis and test methodology.

Using the large-scale powered models and small-scale models, tests verified propulsion component performance and propulsion elements in key areas such as the inlet lift systems and exhaust systems. Low Speed

FLT DECK
LHA-3

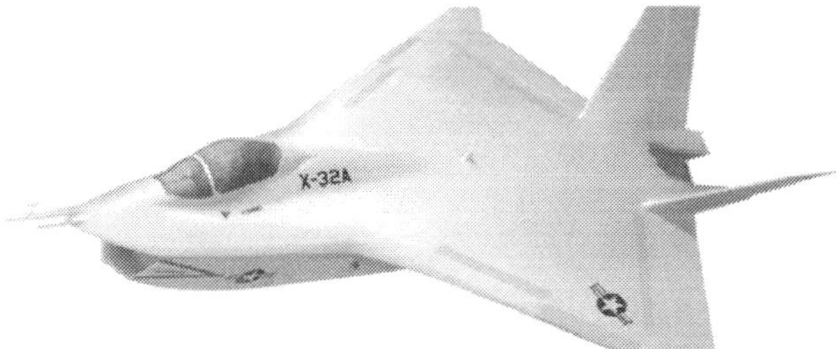
X-32A

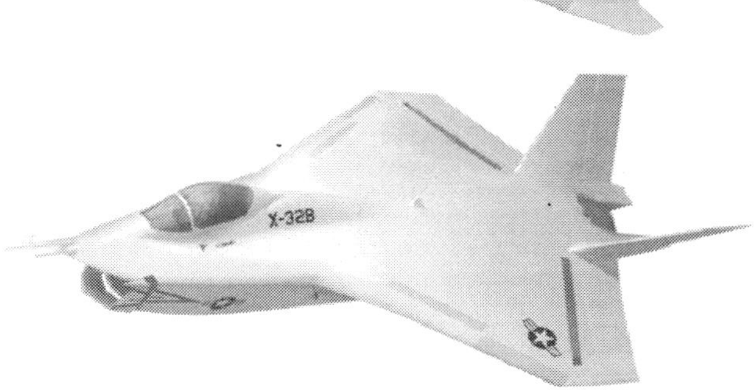
X-32B

Phantom Works

Previous page, top and three left: The Drawings released of Boeings JSF in the various stages of the Concept Evaluation program. Previous page, three right: The X-32 during final assembly. All Boeing **Above: The X-32A on the ramp at Palmdale, California, clearly shows the aircraft's extremely large wing which was the aircraft's main outstanding characteristic.** Boeing

The structurally complete Boeing X-32A CDA is towed from the assembly hanger at Palmdale to be weighed in 1999. Boeing

Aerodynamics studies involved wind-tunnel tests to develop configurations and to establish the stability and control data to determine the configuration's flying qualities. The tests were conducted in three separate low-speed wind tunnels. High Speed Aerodynamics studies focused on transonic and supersonic flight. These tests, which were conducted in six different wind-tunnel facilities in the United States and United Kingdom, further refined the configuration and verified vehicle performance. High Angle of Attack tests were performed at low speeds up to 90 degrees angle of attack and up to 50 degrees angle of attack at high speeds. The tests were conducted using sub-scale models in four different wind tunnels. Other tests included Ground Effects and weapons integration. The former involved an extensive series of tests (including those with the LSPM) involving a combination of small-scale and large-scale models to determine hover, transition effects and hot-gas ingestion. They involved aircraft and weapon models to measure weapon bay acoustic characteristics and confirm internal weapon bay separation predictions.

To say the Boeing X-32 was unorthodox compared with other modern fighter designs is perhaps a bit of an understatement. The aircraft is certainly no beauty contest winner; however, it was designed not to look good, but to get the required job done. The main characteristics of the X-32 are the large one-piece trapezoidal wing, which is of composite construction. The slab sided forward fuselage has a broad distinctive chin mounted inlet at front. An advantage of the deep fuselage was the large internal volume to accommodate side mounted weapon bays, which are easy to load for the internal carriage of weapons in keeping with the 'stealth' qualities of the aircraft. The bays are located on the fuselage sides, just under the wing/fuselage junction. Ordnance such as the GPS guided Boeing Joint Direct Attack Munitions (JDAM) and Raytheon AIM-120 AMRAAM could be mounted on ejector racks with very fast cycle times for the bay door actuation to minimise the time the aircraft could be exposed to excess radar returns. Boeing designed its JSF with side-mounted weapons bays for both operational and supportability reasons. The side-mounted bays allow the pilot to open the bay away from enemy radar and drop a weapon without compromising the JSF's low observable characteristics. The design also benefits the ground crews by allowing eye-level access to the weapons bays for maintenance and weapons loading. The operational JSF would have been able to internally carry two air-to-ground weapons and 2 AIM-120 AMRAAM's as a baseline load, as well as a full range of other conventional weapons.

The slab sided fuselage was divided into two sections consisting of the fore-body and the under-wing section. The front section of the X-32, which was designed in Seattle, but built in St Louis, contains the cockpit and avionics including radar in the proposed production

Above: Certainly no beauty contest winner, the X-32 profile was dominated by the large slab sided fuselage, which accommodated the internal weapon bays. Boeing

variant. The bulk of the large under-wing section was designed and built in Seattle and the wing-box was built in Palmdale. The starting point for the X-32 wing structure was advanced sine-wave technology developed for the F-22 by Boeing. The wing, which in addition to being designed for a large fuel volume is optimised for low-weight, low-cost construction and a high degree of battle damage tolerance. The Boeing wing offered good high alpha characteristics. It was a multi-spar structure, which supported single-piece upper and lower composite skins constructed of thermoplastic matrix.

Wing sweep was a consideration of aerodynamics, radar signature and avionics, as in production aircraft many sensors would have been buried in the leading edge. The large one-piece wing remained virtually unchanged since the earliest configuration AVX-70 of 1993 was defined for the then ASTOVL program. Through to the configuration 350 of 1995, the wing retained tip-mounted fins and rudders, which doubled as winglets. The initial JIRD 1 studies in 1995, focused on size, speed and stealth. It was indicated that twin vertical tails were lighter, resulting in a shape, which evolved as the X-32. The vertical tails were canted outward, providing stability and control during aircraft carrier approaches at high angle-of-attack. Introducing the cant allowed full aerodynamic control to be retained even if the 2-dimensional pitch-vectoring exhaust nozzle failed, which along with the other control surfaces was fully integrated into the flight-control system. The wing is the central part of the aircraft. According to Boeing "We basically designed it and attached a fuselage to it." The large wing, which is structurally deep, can house up to 8600-kg (19,000-lb) of fuel allowing an aircraft around the size of an F-16 to carry three times more internal fuel.

The main difference between the X-32A and the X-32B was the latter's direct lift propulsion system and smaller span wing, which lacked the leading edge control surfaces of the X-32A wing. The X-32A had a wingspan of 36-ft while the shorter wing of the X-32B spanned only 30-ft. Although the model 375 proposed PWSC STOVL variant was slightly shorter than the CTOL and CV variants, in the X-32 CDP, the X-32A and X-32B are the same length at 44-ft 8-in (without data probe). The published weight for the X-32A is 36,800-lb Max TOGW. There is no released figure for the X-32B. Both aircraft have published maximum speeds of Mach 1.0 although in flight-testing this was exceeded by both aircraft.

Boeing adopted the chin intake as the most affordable solution, as only one would be required and it was less complex and, therefore, easier to manufacture. A

JSF
Phantom Works

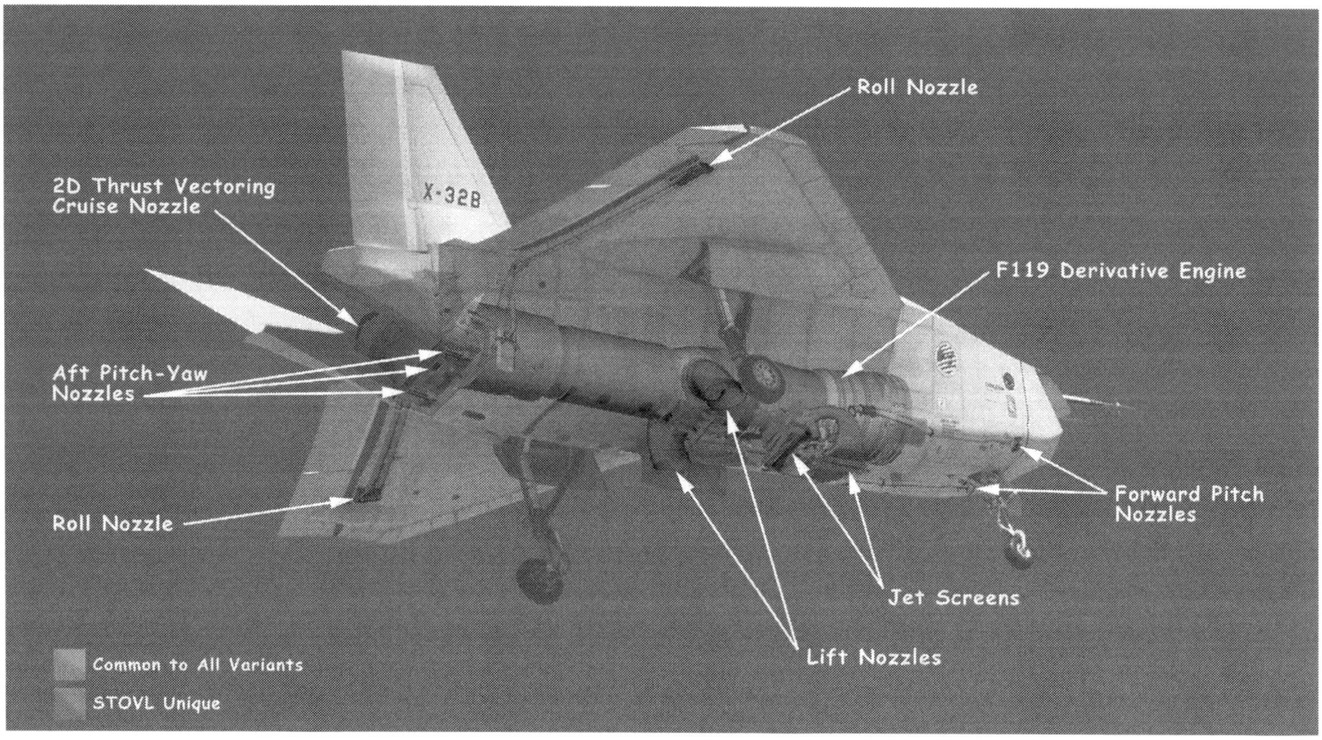

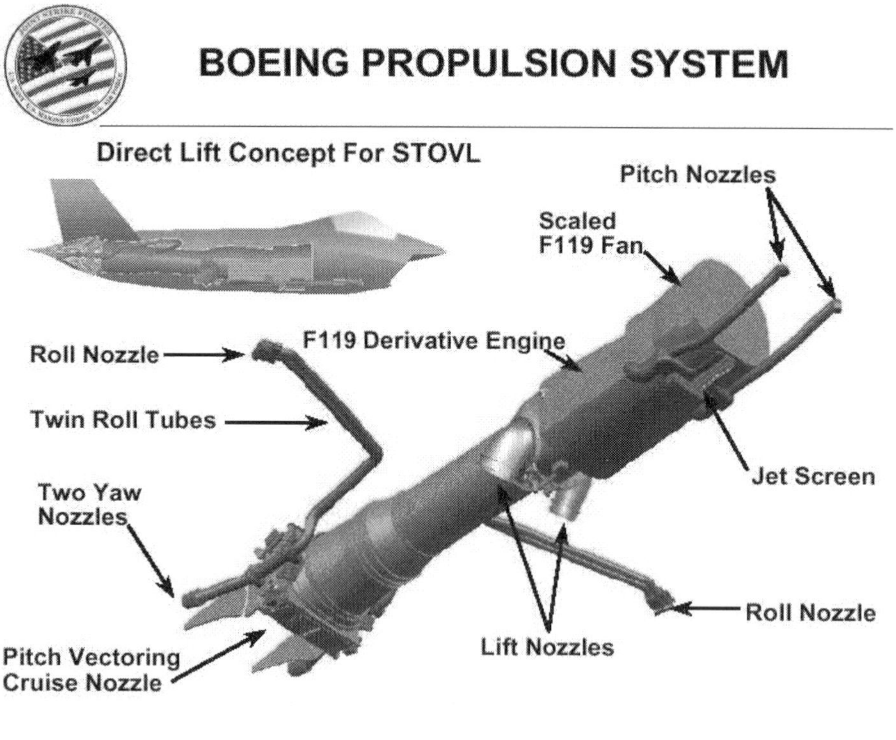

Previous page top: The STOVL propulsion system for the X-32B undergoing ground testing. Previous page bottom: The JSF119-614 engine is prepared for installation into the X-32B. Both Boeing

This page: Two diagrams showing the details of the STOVL propulsion system for the Boeing JSF. Boeing

The Mk16A ejection seat for the X-32 undergoing sled testing. Boeing

challenge, especially for the PWSC, was how to adapt the sharp intake shape to accommodate supersonic flow conditions and low speed hovering conditions. The design team looked at blow in doors on the side and auxiliary inlet doors on top of the aircraft. Another solution studied was fences and systems that would alter shape. After extensive testing including wind tunnel testing on the blow out doors Boeing adopted the translating cowl as the most affordable solution. During slow speed operations or in the hover, when the engine requires a smooth subsonic flow of air, the translating cowl would move forward and back on rail mounted actuators revealing a blunt lip.

The chin mounted intake incorporates a radar blocking device, which essentially consists of inlet guide vanes, which open fully at maximum power during take-off and hovering (in the STOVL variant), then close down in cruise flight, thus maximising the stealth qualities of the intake design.

In order to accommodate the direct lift propulsion system the engine had to be housed in the forward fuselage to allow the main thrust nozzles to be at the aircraft's centre of gravity. This, combined with the requirement to fit existing aircraft carrier lifts, resulted in a short nose profile for the X-32A and the PWSC.

To meet its STOVL goals Boeing adopted a direct lift system tried and proven in the successful BAE Systems Harrier, Sea Harrier and later Boeing/BAE Harrier II. Like the X-35, the X-32 CDA use modified variants of the P&W F119 afterburning turbofan engine developed for the Lockheed Martin Boeing F-22 Raptor. One of the key requirements for the F-22 was the ability to supercruise, resulting in the F119 having a large core and a by-pass ratio of 0.2.1. Early in the JAST program the competitors selected the F119 as the basis for the engine for their respective CDA.

To meet the Boeing direct lift power requirements, P&W developed the JSF119-614. As direct lift alone would require a more powerful engine than the X-35B's fan assisted engine, the JSF119-614 was being developed as the most powerful fighter engine in the world. During testing, the -614 reached 50,000-lb thrust, although 42,000-lb thrust would be more normal for flight operations of the CDA. Among the main differences between the standard F119 of the F -22 and the JSF119-614 is the latter 1.14-m (45-in) diameter fan and two-stage low-pressure turbine, which replaced the single-stage unit in the standard F119. The JSF119 retained the modified three-stage fan, six-stage high-pressure compressor (HPC) and single high-pressure turbine of the standard F119 engine. Incorporation of the larger fan was necessary, as a much higher by-pass ratio was required by the fan to feed the aircraft's vertical lift system.

To perform STOVL flight, the Boeing system closes the rear 2-dimensional exhaust nozzle as a pair of butterfly valves open into the airflow-directing engine thrust downward through two-vectoring lift nozzles. Fan duct flow is discharged through a jet-screen, which is aligned laterally and longitudinally beneath the forward fuselage. Additional fan air is directed out to wingtip nozzles for roll control and to the aft fuselage nozzles for pitch and yaw control.

The Boeing X-32A CDA sits on the ramp at Palmdale in late 1999. The aircraft's bulky appearance is emphasised here with the elevated view of the one piece wing, which housed the large fuel volume carried. Boeing

Hot gas re-ingestion has been a major problem for many STOVL designs. To counter this problem, Boeing introduced the jet-screen. This takes cool air from two butterfly valves in the aft fan duct and directs the airflow downwards to form a sort of shield between the X-32 intake lip and the hot gases being expelled from the two lift nozzles, which are positioned at the aircraft's centre of gravity. The lift nozzles can be vectored from 45-degrees aft to 15-degrees forward. The 2-D main exhaust nozzle can be vectored +/-20-degrees when the aircraft is flying in conventional (non-STOVL) mode. The nozzle is articulated around a hinge, which is covered by a flexible hinge cover. Boeing originally could not get the 2-D nozzle to close properly, but this was overcome by incorporation of a P&W developed crushable seal. The requirement was for transition from cruise to STOVL mode to be completed in three seconds. However, Boeing pointed out that transitions had been completed in around 1 second.

Roll's Royce designed and built most of the lift system, including the main lift module, nozzles, roll and yaw ducts and nozzles. When the aircraft is in the hover mode the engine produces some 35-36,000-lb thrust. Around two thirds of the thrust is generated via the lift nozzles and the remaining thrust via the jet-screen and small effectors. For conventional flight, the lift nozzles

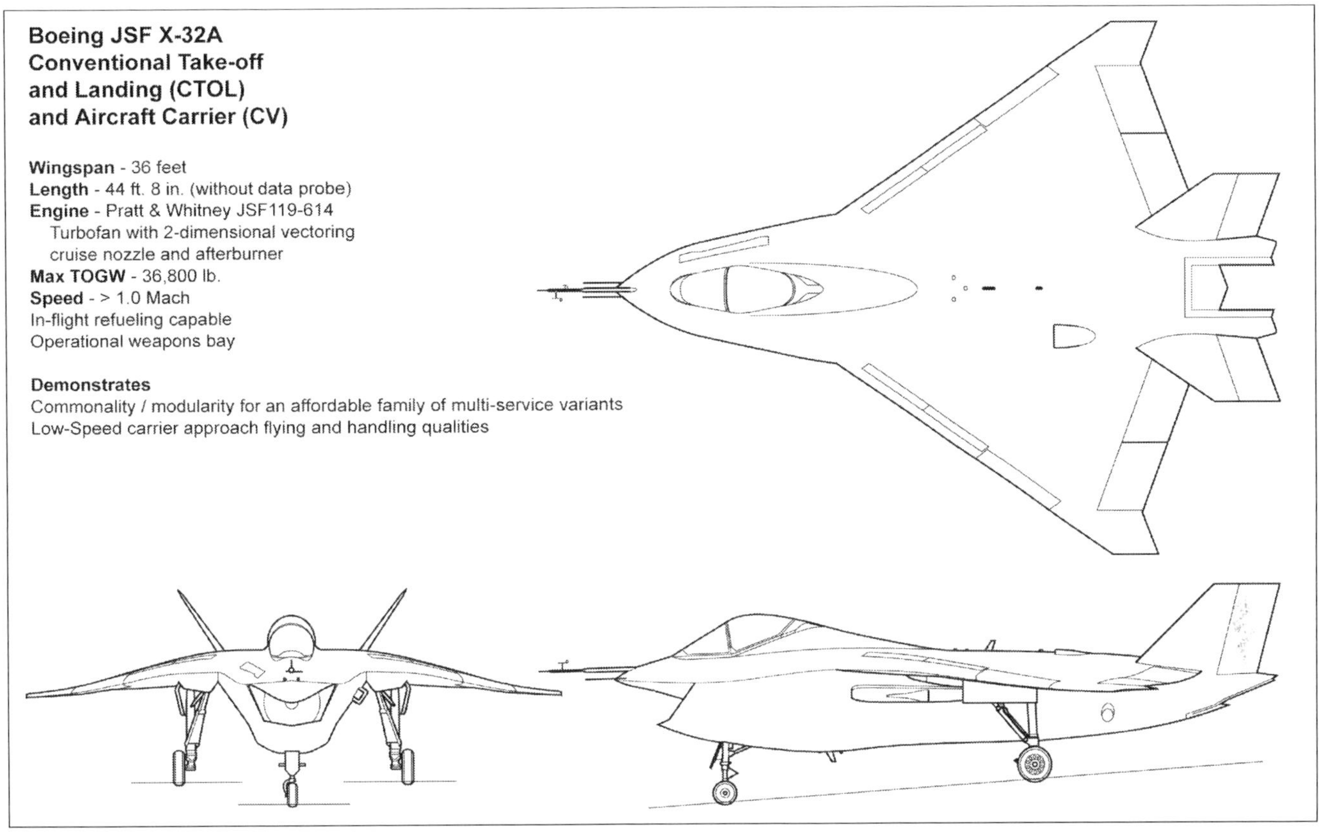

Above: A general arrangement drawing of the Boeing X-32A CDA. Boeing

are closed and thrust flows rearward through the two-dimensional thrust-vectoring cruise nozzle - the same as in the X-32A - to propel the aircraft forward and to supersonic speeds.

The required high thrust required for the STOVL variant benefits the CV and CTOL variants as the extra power increases speed, manoeuvrability and range according to P&W. The lift system adds only an additional 270-kg to the weight.

Boeing cut X-32 tooling costs by more than half over tooling costs in the YF-22 program, with 80-percent fewer defects. X-32 tooling requirements were reduced by 75-percent over the requirements for the YF-22 prototype. Overall fabrication and assembly costs were 30 to 40-percent below projections. Lean manufacturing and other process improvements streamlined construction of the X-32 concept demonstrators. During X-32 assembly, Boeing achieved its goal of less than three defects per 1,000 parts.

Although the Boeing concept stressed affordability, it did not sacrifice performance fundamental to mission effectiveness. The Boeing JSF had stated manoeuvre capabilities equal to or better than the Lockheed Martin F-16 and Boeing F/A-18 and a combat radius and payload significantly better than both these aircraft. The X-32 CDA could carry more than 15,000-lb internal fuel, plus more than 14,000-18,000-lb payload. Combat radius with internal fuel plus internal air-to-air and air-to-ground ordnance met or exceeded government target values for all service variants.

In August 1997, Boeing announced successful completion of two significant propulsion tests. The first test, conducted at the Boeing Nozzle Test facility in Seattle, involved the direct-lift system components, which, when combined with the P&W SE614 engine, comprised the propulsion system for the X-32B STOVL aircraft. Testing of the lift components spanned three months and met the program performance objectives for verifying successful STOVL capabilities for the X-32 propulsion system. Designed and built by Rolls Royce, the lift components - specifically the lift module and spool duct - connected the JSF gas generator to the 2-D cruise nozzle, and converted the engine's horizontal thrust to vertical thrust for both vertical landings and short take-off. The scale model was used to assess the performance and operability of the lift module and spool duct during conventional flight, STOVL operations, and transition from one flight mode to the other. The full range of JSF nozzle pressure ratios, mass flows and lift module positions were also evaluated.

The second series of tests, conducted at the AEDC in Tullahoma, Tennessee, showcased the performance of the high-speed inlet/fore-body compression system for

Boeing JSF X-32B
Short Take-off and
Vertical Landing (STOVL)

Wingspan - 30 feet
Length - 44 ft. 8 in. (without data probe)
Engine - Pratt & Whitney JSF119-614 Turbofan
with 2-dimensional vectoring cruise nozzle and afterburner
STOVL Propulsion system - lift module with two
rotatable lift nozzles, Jet Screen, and fan air supplied
Attitude Control System
Speed - > 1.0 Mach
In-flight refueling capable

Demonstrates:
Commonality / modularity for an affordable family
of multi-service variants STO, VL, Hover and Transition

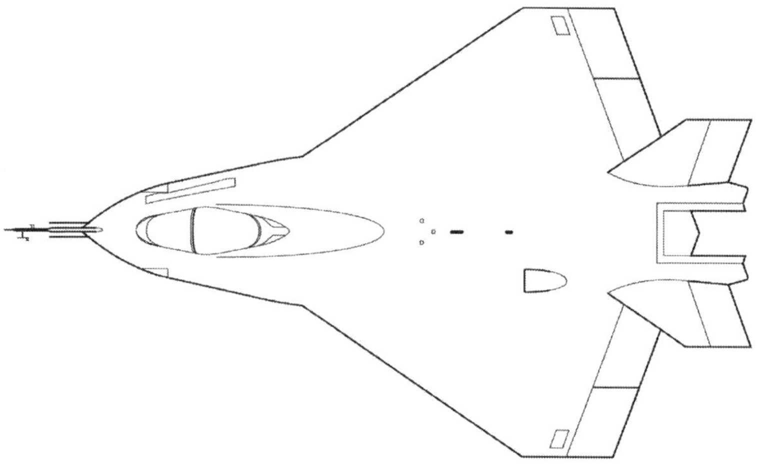

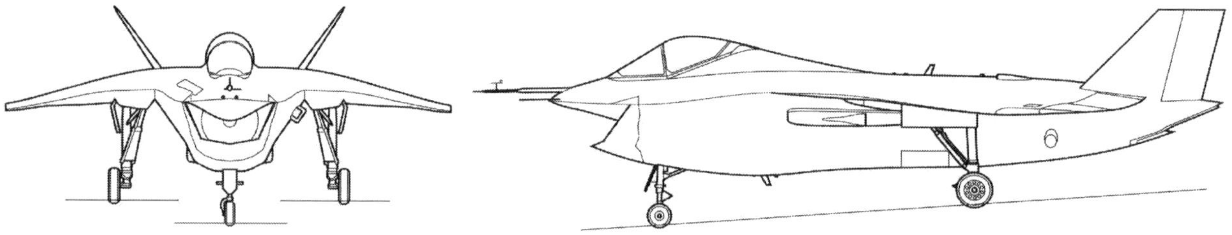

Above: A general arrangement drawing of the Boeing X-32B CDA. Boeing

the Boeing JSF, which during flight provided air to the engine, enabling it to generate thrust to power the aircraft. The three-week-long test employed a 13% scale model and encompassed the full range of JSF flight speeds and attitudes. During approximately 211 hours of testing, actual test speeds included subsonic, transonic and supersonic rates, with model attitudes ranging from 20-degrees nose down, to 60-degrees nose up, and up to 25-degrees of side-slip.

The high-performance capabilities afforded by the wind tunnel at AEDC also enabled the Boeing JSF team to test the high-speed inlet in a variety of conditions including simulated takeoff operations in the low-speed position, used to achieve increased engine thrust during vertical flight operations. Data gathered during testing was critical to defining the operational extremes of the aircraft/engine combination including component compatibility and the optimum speeds at which they can operate and deliver precision performance.

In September 1997, the Boeing JSF program successfully passed its CDA Initial Design Review (IDR), paving the way for fabrication and assembly of the two X-32 aircraft. The combined IDR and Concept Demonstration Aircraft Contract Program Management Review (PMR) addressed significant aspects of the X-32 design including airframe and subsystem reviews, air vehicle weight, propulsion system flight integration and the PWSC, manufacturing planning, flight simulation planning, and first flight elements. The IDR and PMR also highlighted several significant program achievements including air vehicle weight, which was well below the weight Boeing committed to.

On 5 February 1998, Boeing announced that final assembly and first flight of its 2 X-32 CDA would take place in Palmdale, California, with subsequent landings at Edward's AFB, California. The Boeing JSF program used existing facilities in Palmdale to provide the infrastructure and personnel for sub-assembly, final assembly, engineering support, ground testing and preparation for pre-flight checkout and first flight operations. Overall program responsibility continued to be managed from Seattle.

The processes and technology used in the fabrication of X-32 components demonstrated the Boeing JSF team's leverage of best commercial practices. In this case, ones used to develop the Boeing 737 and 777 airliners. Boeing began fabrication of mid-fuselage components for the X-32A in Seattle on 19 November 1997, and structural assembly of the X-32A began a month ahead of schedule on 8 July 1998. Assembly of the X-32A's centre-fuselage started on 21 August 1998 and the first composite wing skin for the X-32A, which measured almost 29-ft across, was flown to the final assembly plant in Palmdale on 4 November 1998. On

Media
VSTOL

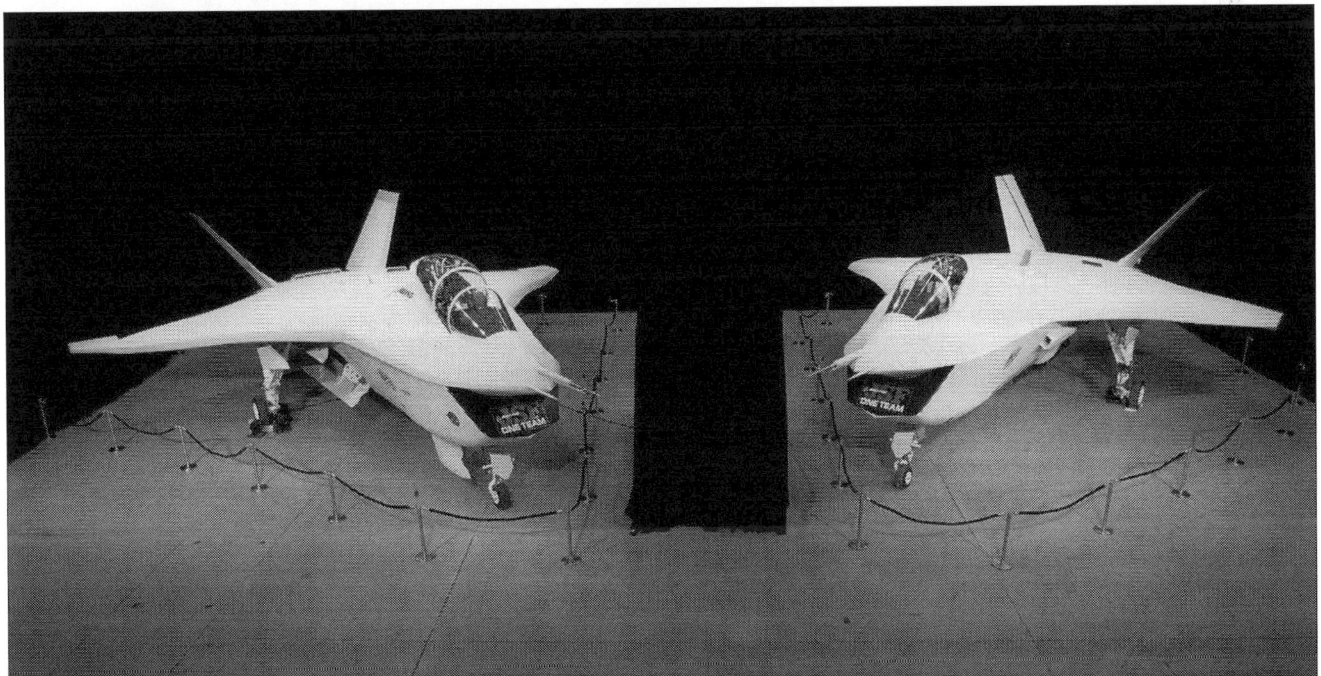

X-32A

Previous page top: The X-32A cockpit. Previous page bottom: The X-32A is seen during engine run trials prior to its first flight. Both Boeing **Above: Boeing demonstrated the avionics for its JSF candidate in its Boeing 737 Avionics Flying Laboratory.** Boeing

26 March 1999, Boeing completed manufacture of the forward fuselage component of the X-32A, which was then transferred to the Boeing X-Vehicle facility at Palmdale for final assembly. On 23 September 1998, Boeing began assembling the forward fuselage of its second JSF CDA, the X-32B, more than two months ahead of schedule.

On 22 June 1998, P&W began ground testing the second of two F119-SE-614 developmental engines. The F119-SE-614 also benefited from the 6,000-hours of ground testing accomplished on the standard F119 which powers the F-22 Raptor. Tests to validate the performance and operational response of the engine were conducted at P&W's West Palm Beach, Florida facility. Initial sea level testing of the Boeing CTOL engine continued for approximately one month. Afterward, the engine underwent altitude testing at the AEDC in Tullahoma, Tennessee. In November 1998, Boeing Completed the Final Design Review for the X-32.

On 8 July 1999, Boeing announced it had successfully completed the first test phase of the vehicle management system (VMS) for its X-32. The VMS operates the flight controls, including the actuators and hydraulics, the environmental control system and other subsystems. Boeing used a full-scale actuator test rig that simulated the aircraft's hydraulic and actuation systems. Both engine-driven pumps and auxiliary power unit-driven pumps were used to power the hydraulic systems. A few weeks later on 30 July, Boeing connected electrical power to the X-32A. The cockpit's interior lighting, multifunction displays, heaters and several display panels were the first systems powered by an external source. Each system tested was fully operational. A few weeks later on 20 August, the X-32A was towed from its final assembly position to a structural test rig in a hanger at Palmdale for structural proof testing. This move signalled the end of final assembly of the aircraft and the beginning of the ground test phase. Temporary landing gear was used for the move. A test fixture with more than 100 hydraulic actuators applied stresses to different parts of the aircraft to simulate in-flight aerodynamic pressures and carrier landings. The tests confirmed the integrity of the aircraft's structural design by simulating flight and landing conditions. No structural problems were encountered, and critical strain measurements during the test matched engineers' predictions. Structural proof testing was completed on 9 October 1999. Assembly of the X-32B continued and the single-piece wing was attached to its fuselage on 20 September 1999. Overall, assembly of the wing for the X-32B took one-third less time than for the X-32A. In early November 1999, P&W completed final assembly of its JSF119-614 flight-test engine and in late November 1999, The Boeing JSF One Team completed testing, which verified that the fuel system developed by a team led by Flight Refuelling Limited for the X-32A was safe, effective, leak-free and ready for flight-testing

Previous page: Boeing displayed a full-scale mock-up of its revised PWSC (Preferred Weapon System Concept) at the Farnborough show in July 2000. Boeing

This page above: The Boeing JSF PWSC full-scale mock-up sits on the ramp with a production Boeing F/A-18F Super Hornet at Farnborough in July 2000. Boeing

The Boeing X-32 one-piece wing contained all the fuel tanks, a design that helped ensure leak-free performance. The X-32A also continued final systems installation and check out in preparation for flight testing in 2000.

On 14 December 1999, Boeing unveiled both its X-32A and X-32B CDA at Palmdale. Thousands of employees, suppliers and customers around the world watched as Boeing surprised observers by introducing its X-32B model along with the X-32A. On 6 March 2000, P&W delivered the JSF119-614 CTOL flight test engine and in April the engine was installed in the X-32A. Designated the YF004, the engine had successfully completed 45 hours of acceptance testing at P&W (UTX) facility in West Palm Beach, Florida. The YF004 engine met or exceeded requirements in all categories, including available thrust levels and engine operability, building upon more than 900 hours of sea level and altitude development testing at P&W, and at the AEDC. In May, P&W completed flight clearance for both the JSF119-614 (Boeing X-32A CTOL) and the JSF119-611 (Lockheed Martin X-35A CTOL and X-35C CV) respectively. The engines underwent performance-

Above: A model of the Boeing STOVL variant PWSC in Royal Navy Fleet Air Arm marking. Author

testing at all power ranges within their respective full flight envelopes. In July 2000, Boeing engineers installed the propulsion system for the X-32B in less than four hours.

X-32A

Wingspan: 36-ft
Length: 44-ft 8-in (without data probe)
Engine: P&W JSF119-614 afterburning turbofan with 2-D vectoring cruise nozzle.
Max TOGW: 36,800-lb.
Speed Mach 1.0
Demonstrated commonality/modularity for a family of multi-service variants, low speed carrier approach, flying and handling qualities.

X-32B

Wingspan 30-ft
Length: 44-ft 8-in (without data probe)
Engine: P&W JSF119-614 afterburning turbofan with 2-D vectoring cruise nozzle.
STOVL propulsion system: Lift module with two rotating lift nozzles, jet screen, and fan supplied attitude control system.
Speed: Mach 1.0
Demonstrated commonality/modularity for a family of multi-service variants, low speed carrier approach, flying and handling qualities. STO, Hover and Transition.

In late 1998, Boeing revised its JSF PWSC design to meet the changing requirements of JIRD 3. The main reason for the changes was that the previous design configuration 372 was too heavy, resulting in design configuration 373, which was considered to be the 90% solution. The decision to change the design from a delta to an aft swept wing with separate horizontal tails was taken over a three-month period in late 1998. However, Boeing did not publicly reveal the new configuration until February 1999.

The changes saw the Boeing PWSC metamorphous to a smaller swept wing, with a separate horizontal tail moving away from the previous delta wing concept. As well as reducing weight, the switch was made to increase pitch control and improve carrier suitability for the CV variant. According to Boeing, this was not possible with the delta design without reducing low observable performance. The inlet was also redesigned from forward swept to an aft swept lip, which reduced weight and improved high angle-of-attack performance, but did not degrade the aircraft's LO performance. The trailing edge of the wing is raked back at 20-degree as opposed to forward 20-degree as in the previous design. To further reduce weight, the twin outward canted vertical tails were redesigned, as was the canopy. The aircraft's fuselage remained more or less unchanged from the earlier configuration 373, with the 55-degree sweep remaining. The powerplant remained the same as well. The STOVL variant was configured with a clipped 9.15-m (30-ft) span wing, while the span of the CTOL variant was 11-m. The STOVL variant in the new PWSC design was also proportionately shorter in length than the CTOL variant. Length for the CTOL variant was 14.42-m, while the STOVL variant was 14.03-m.

Above: The Boeing PWSC simulator showing the large format display screens, which characterised the JSF cockpits. Boeing

The reduced length of the STOVL variant was accomplished by removing a section of the aft fuselage behind the trailing edge. This resulted in a 136-kg weight saving, although Boeing hoped to save the weight elsewhere, allowing it to restore all PWSC variants to a common length for production aircraft. The height of the CV variant of the PWSC would be 4.05-m; 3-cm taller than the CTOL and STOVL variants as a result of the CV having a stronger undercarriage with a double nose wheel.

The USAF conventional take-off and landing variant would have had an internal mounting for the 27-mm cannon, while the USN carrier (CV) variant would have had a reinforced structure, stronger undercarriage mentioned above and a tail-hook for the aircraft carrier role. Like the X-32 concept demonstrators, the STOVL variant of the PWSC would also differ from the CTOL and CV variants in having the plumbing necessary for the STOVL mission.

Avionics Flying Laboratory

On March 26, 1999, a highly modified Boeing 737-200 made its first flight from Boeing facilities in Wichita, Kansas and began its service as a flying laboratory in support of the Boeing JSF program. After two test flights to evaluate the aircraft's handling characteristics and verify instrumentation and flight data collection systems, each lasting about two hours, the Boeing 737 AFL (Avionics Flying Laboratory) was delivered to Boeing in Seattle. In a less-than-one-year design and modification program, Boeing fitted a 48-inch nose and assembly to the forward pressure bulkhead. The elongated nose housed avionics including a development AESA (Active Electronically Scanned Array) radar and instrumentation to aid development of the Boeing JSF PWSC. The aircraft was also fitted with several antennas, a heat exchanger and provisions for a supplemental power system. Boeing initiated engineering design for the program in April 1998, with modifications to the aircraft beginning in December that year.

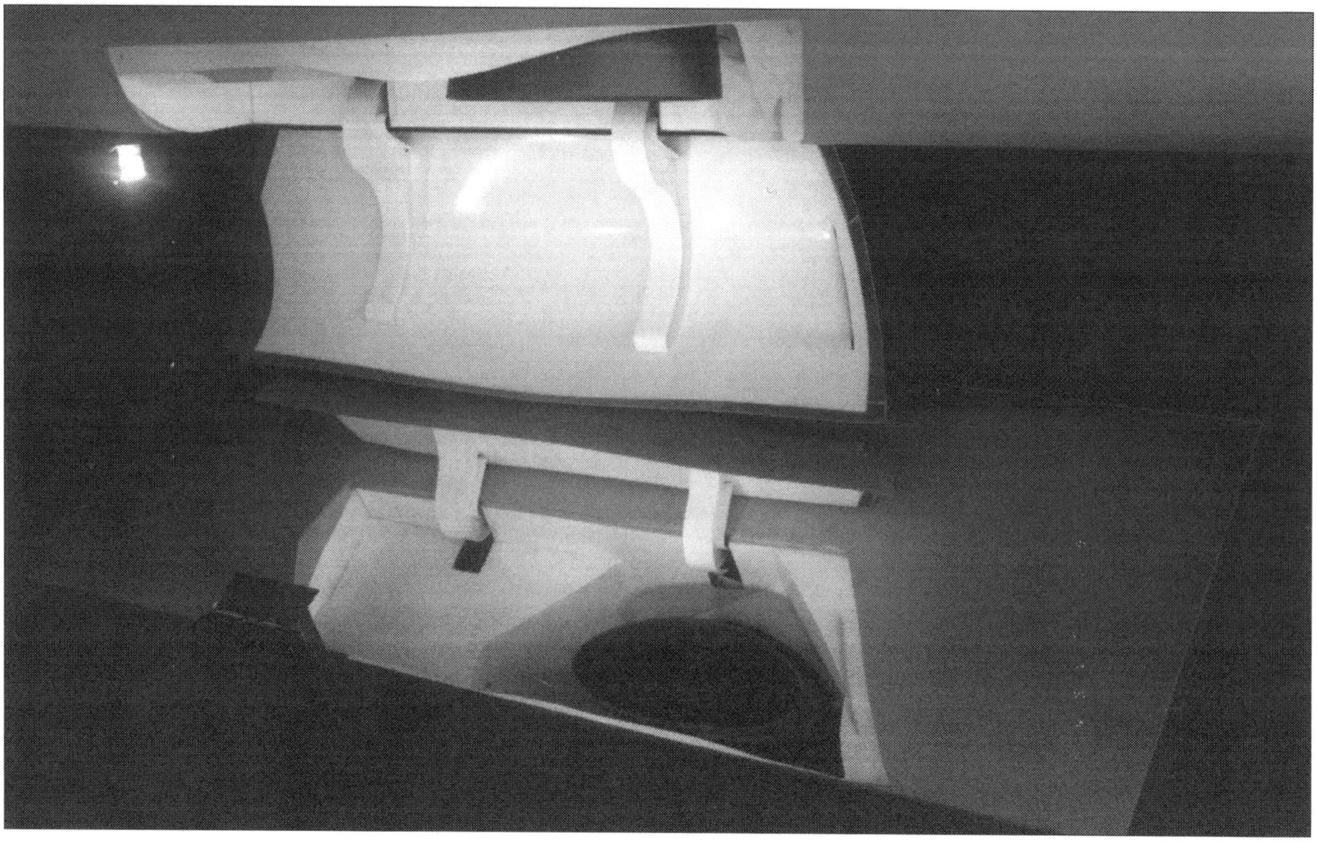

Top: The 2-D thrust-vectoring main nozzle and bottom the main lift nozzle of the Boeing JSF. Both Author

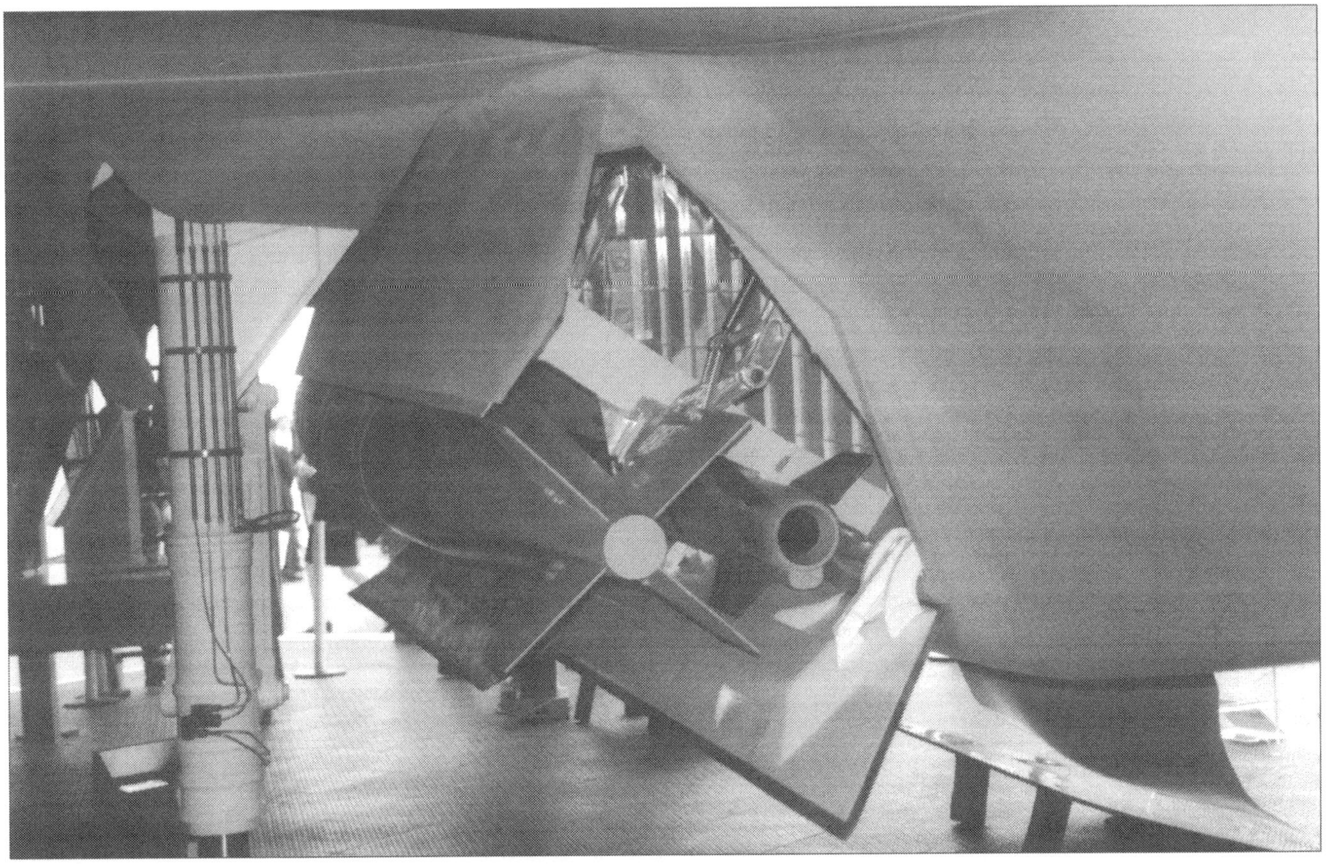

Previous page top: The JSF was designed as a stealthy platform, and in keeping with this characteristic the aircraft were designed with provision for internal weapons carriage. The baseline load was two Boeing Joint Direct Attack Munitions and 2 AIM-120C AMRAAM air to air missiles. One of each is seen here in the Boeing mock-up.
Previous page bottom: When stealth is not a priority, the JSF can carry ordnance on external weapons stations. Here, the Boeing mock-up is displayed at Le Bourget, Paris in June 2001, with an external load of JDAM, Raytheon Joint Stand-Off Weapon and Matra BAe Dynamics Alenia Advanced Short-Range Air to Air Missile on a stub pylon. Both Author

This page above: This ghosted view of a model of the Boeing PWSC shows the internal weapon bays and the stowed undercarriage. Author

On 9 December 1999, Boeing began five months of integrated avionics testing with the first test flight of the AFL, which would significantly reduce technical risk in JSF avionics development by letting engineers evaluate and troubleshoot the avionics systems before they were installed on the JSF.

A representative JSF cockpit was installed in the aircraft's cabin. The AFL carried a crew of avionics engineers, who monitored test data, fixed problems and directed testing to take advantage of emerging results. With near instantaneous fusion of data from on-board and off-board systems, including other aircraft, the Boeing JSF avionics system freed the pilot from actively managing data flow in the cockpit. Key components tested included radio frequency and Electro-optical sensors and a prototype core processor. The processor, which runs the mission-system software, used an open architecture that enables reuse and easy portability.

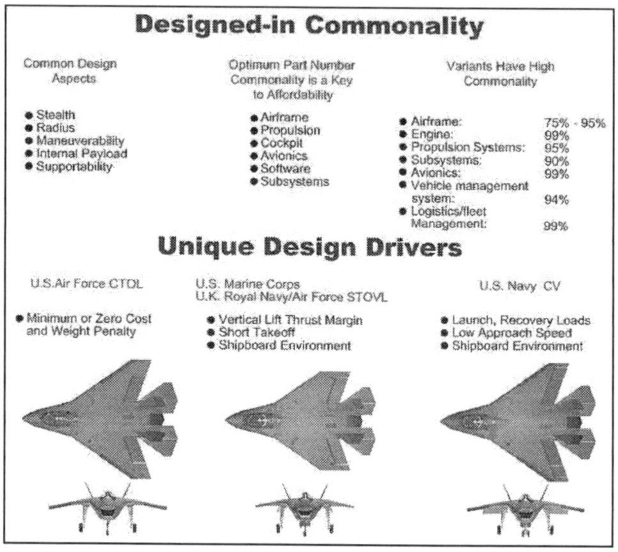

Chart showing the large degree of commonality between the three Boeing JSF variants. Boeing

Boeing Joint Strike Fighter Conventional Take Off and Landing	
Length:	47' 5"
Wingspan:	36' 0"
Height:	13' 4"

These views show the metamorphosis from the delta wing X-32 to the conventional tail-plane PWSC JSF. All Boeing

Above: Artist rendering of the Boeing PWSC operational aircraft. The first aircraft is in clean configuration, while the second aircraft is shown with external stores and internal weapon bays open. Boeing

With hundreds of hours of F-22 avionics testing aboard a Boeing 757 flying test bed under its belt, Boeing had already demonstrated how its test-bed approach could reduce avionics development costs, risk and future flight-test hours. The lab's superior range allowed more thorough testing. It carried more instrumentation than a fighter size aircraft could and doubled as a laboratory on the ground for tests that don't require flight conditions. Unlike a static ground-based lab, the AFL demonstrated JSF capabilities in a dynamic, airborne setting against a wide variety of real targets embedded in their environments.

In May 2000, Boeing successfully demonstrated multi-sensor fusion of its JSF avionics aboard the AFL. The ability to fuse inputs from multiple sensors and display the information in an easy-to-use format for the pilot was a key element of the Boeing JSF integrated avionics. During a series of flights, the AFL's synthetic aperture radar and targeting forward-looking infrared sensors collected target-area information, which was processed by automated target-cueing software to locate and identify the targets. The resulting data was then fused to present the pilot with a composite situational display. Additional target information detected by the electronic warfare sensor was also fused, allowing the pilot to locate and identify the threats without emitting signals. Each function was accomplished in a real-time environment, using low-cost commercial off-the-shelf processors. Combining the data from multiple sensors reduced the number of false alarms a pilot sees during a mission and helps determine where the target or threats are sooner, allowing the pilot to take the necessary "attack" or "evade" steps faster. This rapid decision making process allows the pilot to spend less time over threat areas.

On 28 June 2000, Boeing successfully demonstrated its JSF integrated weapon system's capabilities during a live-fire exercise at the White Sands Missile Range in New Mexico. The Boeing AFL used its JSF mission systems to gather targeting data from off-board systems and fuse it with data gathered from on-board systems. The AFL then provided refined targeting information to a Boeing F-15 Eagle representing a JSF, allowing it to attack a ground target with a GPS-guided JDAM. The AFL's onboard sensors also provided a target damage assessment to both airborne and ground-based exercise participants.

Above: The Boeing PWSC full-scale mock-up outside the Boeing plant. Boeing

The JSF system's ability to obtain information via off-board systems allowed the pilot to locate and attack enemy targets without emitting signals, making the aircraft less vulnerable to detection. In the live-fire exercise, the F-15 did not emit any signals since its target data came from the AFL. During the exercise, the AFL used its Joint Tactical Information System (JTIDS) to send and receive information in Link-16 data format. JTIDS capability enabled the AFL to receive off-board data - air tracks and synthetic aperture radar maps - process it, and then transmit the data package to the other exercise participants. The Boeing JSF open system architecture allowed easy integration of the Link-16 capability into the AFL's JSF mission systems suite. The exercise also demonstrated automatic target cueing, image registration and battle damage assessment using AFL sensors. Boeing said that "No one has successfully put all of these capabilities together in a single platform before."

By July 2000, the AFL had flown around 50 missions amassing well in excess of 100 flight hours and had entered the final phase of testing, which was concluded a few weeks later. This final phase focused on integrating the Raytheon Synthetic Aperture Radar (SAR) to the Advanced Targeting Forward Looking InfraRed (ATFLIR) and BAE Systems Helmet Mounted Display (HMD). The AFL entered the final flight phase after an 11-day changeover to a new Raytheon built radar, which was around 70-kg (150-lb) lighter than the first radar tested.

High-fidelity full-scale Model

In order to validate the radar, antenna and stealth performance of its JSF the Boeing One Team built a full-scale high-fidelity model of the PWSC. The model underwent tests at Boeing facilities in Seattle and at White Sands Missile Range, New Mexico. Boeing confirmed that the RCS tests confirmed that the PWSC met service requirements for stealth.

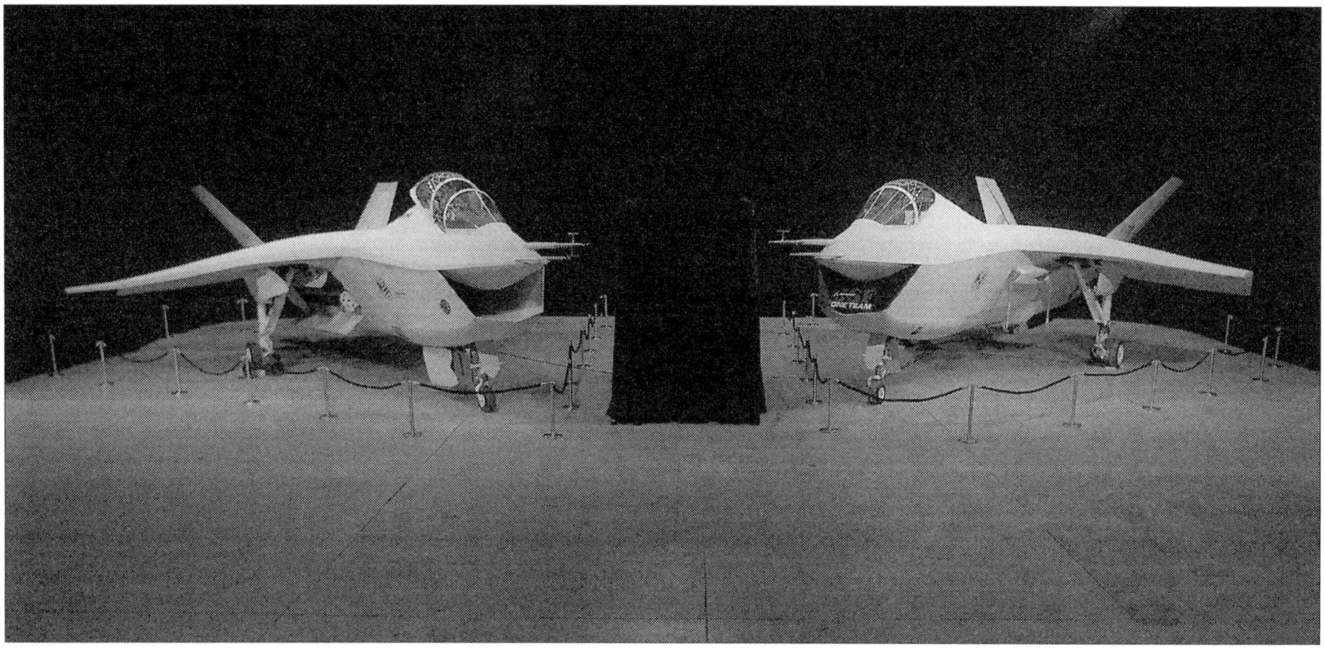

Above: While Boeing undoubtedly proposed a sound PWSC concept, it was the two X-32 CDA what would be the only Boeing JSF design to fly. Boeing

Full- Mission Simulations

Boeing successfully completed four Full Mission Simulations (FMS) for its customer. The simulations helped validate the Boeing PWSC. The FMS incorporated capabilities developed over 3500 of hours of testing, simulations and refinement of the Boeing JSF cockpit, pilot vehicle interface, survivability, flying qualities, weapons integration and mission avionics.

A cockpit demonstrator used in the FMS offered pilots a highly accurate simulation of the aircraft's controls and systems. Boeing JSF capabilities featured in the FMS include Fusion; Information from onboard and off-board sensors is fused into one integrated picture for JSF pilots. Ease of Operation; the demonstrator contains accurate representations of the JSF's controls, advanced intuitive displays and symbology, voice recognition, helmet-mounted display and real-time updates to mission tasks.

The demonstrator simulates the JSF's target location and threat identification capabilities, as well as the accuracy of its weapons. Survivability, Stealth, situational awareness and aircraft performance highlight the Boeing JSF's balanced approach to survivability. In simulation exercises: Pilots from the USAF, USN, USMC and the UK Royal Navy flew approximately 100 simulated air-to-air and air-to-ground missions during a seven-day period, which required them to locate and defeat enemy forces. Boeing linked its JSF full-mission simulator in St. Louis with four high fidelity, networked F-15 operational training devices at Elgin AFB, Florida. The training scenarios, with JSF and F-15 pilots flying together in the same threat environment, demonstrated how air crew at different locations, with different types of aircraft, can practice JSF-representative missions together. In addition to St. Louis, JSF cockpit demonstrators were located in Arlington, Virginia, and Seattle, Washington.

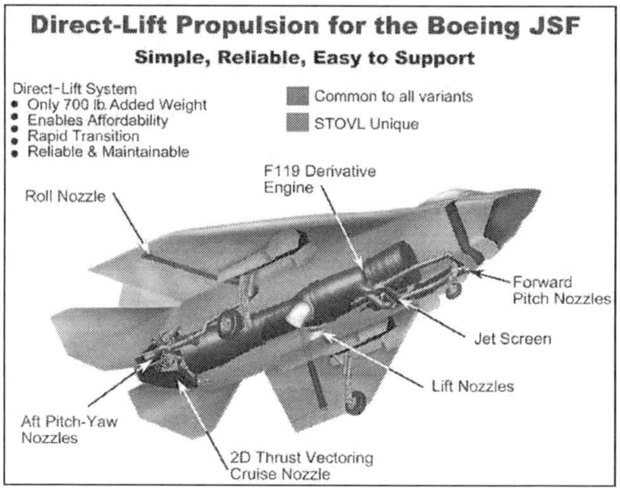

Above: A cutaway drawing showing the STOVL system of the Boeing STOVL PWSC. Boeing

Chapter Three

Lockheed Martin X-35 Concept Demonstrator Aircraft

The Lockheed Martin X-35A CTOL demonstrator is seen during night engine runs at Palmdale in 2000, prior to commencement of the flight test program. USAF

In keeping with one of the fundamental principles of the JSF program, all three X-35 variants were designed with a large degree of commonality. The basic shape of the X-35 was frozen at iteration 230-3 in 1997. It is often said that from a distance the X-35 basic shape with its side mounted intakes and twin tail configuration gives an illusion of the twin-engine F -22 Raptor. While there are similarities in the basic airframe shape, the X-35 is a markedly smaller single-engine design. From the frontal aspect of the X-35, the smooth diverter-less air intake contrasted with the forward angles of the inlet lips. Two serpentine intake ducts concealed the engine face with the twin curved air inlets being designed to provide engine signature line of sight blockage, feeding air via a diverterless duct flight tested on an F-16 during which it was flown at Mach 2 without any moving parts.

The bubble starboard hinging canopy sits ahead of a humped spine designed to create the necessary space for the shaft-driven lift fan installed in the X-35B. The lift-fan is housed in a bay covered at top and bottom by louvered doors with serrated edges, which open and close to control airflow to the fan. From a rear on aspect, the large nozzle for the Pratt & Whitney JSF119 engine was the dominant feature.

This model of the X-35B lift-fan underwent testing in the NASA Lewis Research Centre wind tunnel. NASA

In common with many other experimental aircraft, the X-35 utilised many components from other aircraft, either in service or retired. All three X-35 variants were designed for an 8-g limit and speeds of up to Mach 1.8. The X-35A and X-35B shared the same airframe, with the aircraft being converted from X-35A CTOL to X-35B STOVL configuration at the conclusion of the X-35A flight-test program. Once converted to X-35B the aircraft could be distinguished from the X-35A by the inclusion of the lift-fan system and roll control posts. The X-35C CV variant could be distinguished from the other two variants by its larger span wing and wing surface area due to this variants larger outboard wing sections and control surfaces. All three variants have identical wiring, hydraulic system, environmental control system, secondary power, and cockpit. There are some exceptions, such as flight control actuators on the X-35C, which required larger actuators due to that variants larger horizontal tails. The CV variant also had larger flaperons. The X-35A CTOL variant had no ailerons; therefore, no aileron actuators were required. The X-35 has a completely digital flight control system, in that nothing in the cockpit is attached to anything that moves, such as ailerons.

Like the Boeing X-32, the X-35 CDA was powered by a variant of the P&W F119 turbofan. Among the main differences between the standard F119 and the JSF119-614/611 was the latter's 1.14-m (45-in) diameter fan and two-stage low-pressure turbine, which replaced the single-stage unit in the standard F119. The JSF119 retained the modified three-stage fan, six-stage high-pressure compressor (HPC) and single high-pressure turbine of the standard F119 engine. Incorporation of the larger fan was necessary as a much higher by-pass ratio was required by the fan to feed the X-32's direct lift system and help drive the lift-fan in the X-35.

In addition to using the core and other proven features of the F119 engine, the Pratt & Whitney team capitalised on technology being developed and confirmed in the advanced programs area. These programs included CEASAR (Component and Engine Structural Assessment Research) Program, as well as from the more encompassing DoD/NASA/industry Integrated High Performance Turbine Engine Technology (IHPTET) Program.

Above: A model of the X-35B short take off and vertical landing system undergoing testing. NASA

The X-35B STOVL variant featured a unique propulsion system, in which a drive shaft from the Pratt & Whitney JSF119-611 engine turns a counter-rotating lift fan that produced a cool-air lifting force during STOVL operations. The Rolls-Royce fan, actuated by a clutch that can be engaged at any power setting, worked in concert with an articulating rear duct and under-wing lateral-control nozzles to lift the aircraft with some 40,000-lb of vertical force. Because the fan amplified the engines power, the engine was able to run cooler and with less strain, increasing reliability and extending service life.

The shaft-driven lift fan propulsion system includes a vectoring rear nozzle, wing-mounted roll post ducts that provide lateral control and a counter-rotating fan mounted immediately aft of the cockpit driven by a Pratt & Whitney JSF119-611 engine via a drive shaft. The systems principal components are the shaft-driven lift fan, Pratt & Whitney F119-611 engine, three-bearing rear swivel nozzle, and lateral roll-post ducts. The 18,000-lb thrust from the lift fan combined with around 18,000lb+ from the main engine rear nozzle and additional power from the wing roll posts equated to around 39,000-lb+ thrust for vertical lift. Installed propulsion in the X-35B was around 37,000-lb. The fan itself, which is made of titanium, is 1.27-metres in diameter.

All of the X-32 and X-35 CDA had duel Full Authority Digital Engine Control (FADEC). The X-35B also had movable exhaust vanes in the LPT, which were required due to the dramatically changing angle of attack behind the LPT when the clutch was engaged. Additional air is fed to the engine via an auxiliary air inlet on the aircraft's spine. The clamshell doors for this inlet together with the doors below the lift fan and those, which open beneath the main nozzle to allow it to swivel down for STOVL flight, are controlled by the aircraft's vehicle management system.

In November 2000, it was revealed that Lockheed Martin had redesigned the lift-fan for the X-35B to increase reliability before flight-testing began with "minor changes" to improve fan reliability between two and five times. The changes included switching to materials with better-matched coefficients of thermal expansion.

In April 1997, Allison tested a lift fan exhaust nozzle model in the Lewis Powered Lift Rig in support of the Lockheed Martin JSF program, with the tests completed in mid-July 1997. Testing included various configurations, including pitching vectored angles, which ranged from 15° forward to 60° backward, over a range of nozzle pressure ratios. Nozzle flow rates, thrust, and static pressures were measured for each of the configurations tested.

Having lost as part of the unsuccessful McDonnell Douglas JSF team in November 1997, Northrop Grumman joined Lockheed Martin's JSF team in May 1997, the same month that Lockheed Martin completed the Preliminary Design Review (PDR) of the shaft-driven lift fan. This was followed in June by completion of the initial design review for the X-35 CDA and on the 18th of the month it was announced that British Aerospace had joined the Lockheed Martin team.

Assembly of the X-35A commenced in April 1998, at Lockheed Martin's Skunk Works in Palmdale, California and following the final design review in August that year, a full-scale mock-up of the X-35A was unveiled. The previous month initial testing of the JSF119-611 engine had been completed.

In a bid to offset cost overruns, Lockheed Martin switched the roles of its CDA aircraft under construction in early summer 1999. Under the original plan the X-35A was to be flight-tested then converted to the X-35C. However, the revised plan called for the X-35A to be converted to the X-35B, which it was stated would avoid the additional time and cost of fitting the larger wing control surfaces, fins and tail-plane of the CV naval variant. The second X-35 CDA, which was planned to be the X-35B STOVL variant was now to be completed as the X-35C. While the new approach had a number of benefits, it also had some disadvantages as the CTOL aircraft would now be around 1,800-kg (4,000-lb) overweight due to the lift system hardware.

Once final assembly of the structure, aircraft systems and avionics installations had been completed, the X-35A was moved from assembly tooling to the factory floor at the Skunk Works on 18 September 1999, where additional assembly activities included installation of control surfaces, actuated doors and the landing gear and the JSF119-611 CTOL flight engine was installed in the aircraft on 9 December 1999 with airframe assembly completed by late March 2000 and emerging in June that year, before conducting more tests leading to the first flight.

Two-view general arrangement drawing of the Lockheed Martin JSF design around the time of the CDA award in 1996. The plan view is of the USAF CTOL variant. The top and bottom side views are of the X-35B STOVL variant and the centre side-view is of the USN CV variant. Lockheed Martin

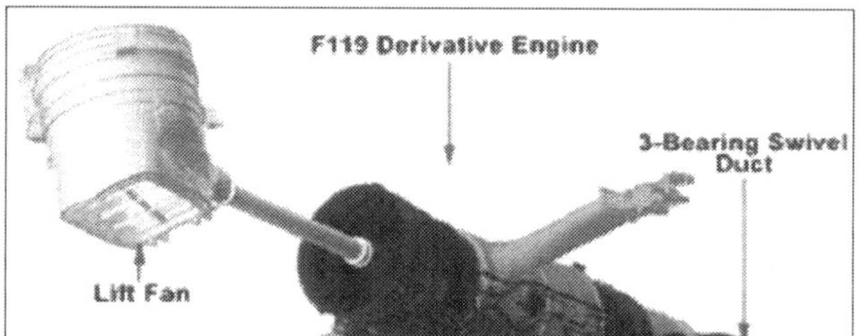

Left: In 1998, Lockheed Martin unveiled this mock-up of its X-35A CDA. LMTAS Above: A diagram of the X-35B STOVL system, which was based on the Pratt & Whitney F119 derivative engine. The other main components of the STOVL system were the lift-fan, 3-bearing swivel duct and the roll control ducts. Pratt & Whitney

Previous page: Much of the technical risk of the Lockheed Martin JSF program centred on the unique STOVL propulsion system developed for the X-35B and later F-35B. The X-35B is seen here with the top doors for the shaft-driven lift-fan in the open position.

This page above: The X-35B STOVL system is put through its paces during ground testing at Palmdale in early 2001.

Next page top: The X-35C could be distinguished from the X-35A/B by its larger span wing. Bottom: While both teams had a common set of requirements, they took different paths from which emerged two totally different looking designs; the X-32 left and the X-35 right. All LMTAS

Above: While the PWSC would have a planned forward hinging canopy, the X-35 CDA were designed and built with a side-hinging unit. LMTAS

Chapter Four

X-32 Flight Test Program

Above: The Boeing X-32A Conventional take-off and landing concept demonstrator aircraft is seen during an early test flight over the desert of southern California. Boeing

In no way was the JSF concept demonstrator aircraft (CDA) flight test phase to be a fly of between the Boeing X-32 and the competing Lockheed Martin X-35 CDA aircraft. This was not to be a program in which the better performing aircraft was to be chosen. Instead, during flight test the competing JSF demonstrator aircraft were required to successfully demonstrate three objectives.

- commonality and modularity among three JSF variants
- low-speed carrier approach flying and handling qualities
- Short take-off, transition, hover and vertical landing.

The Boeing flight-test plan for the X-32A was to complete key customer requirements before proceeding to company "up-and-away" objectives to expand the flight envelope. The road to a successful first flight began on 24 April 2000, when the Boeing JSF One Team successfully completed the first phase of engine runs for the X-32A CTOL variant. During this test phase, the F119-614 engine ran at idle power to verify system integrity, and then at all power settings, from minimum to maximum afterburner. Over a six-day period the engine was powered up eight times, during which all propulsion-system components operated as designed with no anomalies experienced. The team also performed emergency system tests and emergency shutdowns.

Additional engine tests involving the vehicle management system, Electro-magnetic interference and other aircraft performance requirements were conducted prior to low and medium speed taxi tests, which were conducted on 24 May 2000. The taxi tests verified function and integration of aircraft systems, including steering, braking, engine controls and flight-control surfaces while the aircraft was in motion. Data from the taxi-tests was used to verify that the aircraft was ready to proceed to high-speed taxi tests, which were conducted on 15 September 2000. During the high-speed taxi testing the aircraft reached speeds of up to 130-knots, allowing evaluation of the air data sensor as a precursor to flight.

On 18 September 2000, three days after the high speed taxi trials, the X-32A concept demonstrator got air under its wheels when it made its first flight from Palmdale, becoming the first JSF CDA aircraft to fly. The flight, from Palmdale to Edward's Air Force Base began the aircraft's five-month flight-test program to validate the JSF's flying qualities and performance for conventional and aircraft carrier operations. Following a 20-minute, 30-mile hop from Palmdale, the X-32A landed at Edward's AFB, California. Boeing JSF chief test pilot Fred Knox said, "The aeroplane is a pleasure to fly." During the flight, Knox put the X-32A through some initial air-worthiness tests, including flying qualities and sub-systems checkout reaching an altitude of 10,000-ft.

Following a two-day delay, attributed to high winds that exceeded the limits defined by test flight guidelines, the X-32A completed its second flight on 23 September. During this flight, which lasted 50 minutes, Knox again took the aircraft to an altitude of 10,000-ft and attained an air speed of more than 200-mph. During this second flight the X-32A was put through a further series of manoeuvres and tests to establish the X-32A's basic airworthiness.

Top left: The X-32A during low-medium speed taxi trials in May 2000. Centre and bottom left: The X-32A during its maiden flight on 18 September 2000. Top right: The X-32A approaches for landing at Edward's AFB during its maiden flight on 18 September 2000. Above: A side on view of the X-32A as it touches down at Edward's AFB. All Boeing

Above: The Boeing X-32A CDA during its 10th flight. Boeing

Following these early test flights, the program gathered pace and on 13 October the X-32A conducted its 10th flight, during which the aircraft was flown to an altitude of 10,000-ft and continued to validate the aircraft's carrier variant flying qualities.

FCLP (Field Carrier-Landing Practice) tests began at Edward's AFB on 15 November, demonstrating flying and handling qualities during low-speed aircraft carrier approach. USN Commander Phillip Yates, the US government's lead test pilot for the Boeing JSF, and Boeing lead test pilot Fred Knox demonstrated simulated carrier landings using a Fresnel lens on the ground to provide pilot cues during their approaches to a simulated carrier deck outlined on a runway. By this time, including the 30-minute 15 November flight, Boeing had completed 18 flights and 24 percent of the program's overall flight-test objectives.

Boeing had also completed 54 percent of its carrier variant approach objectives, which were completed on 2 December 2000, with the aircraft flying up to five times a day during the previous week. During the CV tests, 97 approaches and 74 touchdowns, as well as numerous "wave-off," throttle transients and integrated test blocks including roll response and speed stability during FCLP tests were accomplished.

Above: The Boeing X-32A CTOL CDA during an early test flight with its under-carriage extended. Boeing

By 4 December, Boeing had completed 50-percent of the government-defined X-32A flight-test objectives and around the middle of the month, Royal Navy Lt. Commander Paul Stone became the first pilot from the United Kingdom and the fifth pilot overall to fly the X-32A when he conducted a 50-minute flight at Edward's.

Top: The unmistakable bulky shape of the Boeing X-32A CDA seen during the aircraft's 15th flight, over southern California. Above: The X-32A crosses the threshold as it approaches for landing at Edward's Air Force Base, California, at the end of its 15th flight. Both Boeing

Above: The large slab sided fuselage of the Boeing X-32 is readily apparent in this view of the aircraft in flight. While the huge wing effectively acted as a huge fuel tank, the deep fuselage housed internal weapon bays allowing the aircraft to meet or exceed stealth targets. Boeing

In-flight refuelling trials with the X-32A commenced on 19 December 2000. Flying at 20,000-ft and more than 270-mph, Yates manoeuvred the X-32A in behind a Boeing KC-10A Extender tanker from the 6th Air Refuelling Squadron (ARS), Travis AFB, California, during the X-32A's 40th flight. This mission involved three contacts between the X-32A and the KC-10. Because the X-32A was acting as both the CTOL and CV version, the KC-10 was required to refuel the X-32A with a drogue and probe. As well as proving the aerial refuelling capability of the aircraft, this part of the test program had significant short-term goals as well. It demonstrated the ability to in-flight refuel, thus, extending mission time, which would prove significant when the X-32B STOVL variant would be flown across the US to the East Coast NAS Patuxent River, Maryland early in 2001.

On 21 December 2000, the X-32A went supersonic for the first time, during envelope expansion tests, meeting one of Boeings "up-and-away" test objectives. During the flight USAF Lt. Colonel Edward Cabrera took the X-32A to 30,000-ft and exceeded Mach 1.0 at 8:30 am Pacific Standard Time on the aircraft's $49^{\text{th fl}}$ flight. "The X-32A approached and exceeded Mach 1 effortlessly, demonstrating significant tactical potential," Cabrera said afterwards.

On 25 January 2001, Boeing successfully completed the X-32A flight tests with a loaded internal weapons bay. Carrying an instrumented Raytheon AIM-120 AMRAAM (Advanced Medium Range Air-to-Air Missile) and a Boeing JDAM (Joint Direct Attack Munitions), the X-32A's weapons bay doors opened and cycled as part of vibration and acoustic testing during the aircraft's 61st flight, validating the design-predictions of the X-32A's weapons carriage environment.

AMC

Previous page top: The Boeing X-32A acted as both the CTOL and the CV demonstrator. Here the aircraft is conducting Field Carrier Landing Practice (FCLP) evaluations required for the CV test program during its 22nd flight. Boeing

Previous page bottom: Seen during its 40th flight, the Boeing X-32A conducted in-flight refuelling trials with an USAF Boeing KC-10A Extender tanker aircraft. The aircraft conducted trials using both the fixed boom for the CTOL and the hose and drogue for the CV and STOVL variant. Boeing

This page above: The X-32A with its starboard weapon bay door cycled in flight. This view shows the serrated edges of the doors, which enhanced the stealth qualities of the aircraft. Boeing

Following conclusion of its 66th flight on 5 February 2001, the X-32A flight test program was successfully concluded, being hailed by Boeing as "one of the most successful flight-test programs in history." Since it first flew on 18 September 2000, the X-32A had completed 50.4 flight hours with six different Boeing and government pilots at the controls. As well as meeting all of the government test objectives, Boeing also provided an impressive array of additional flight test accomplishments beyond those identified by the customer.

Above: The Boeing X-32A CDA approaches the runway at Edward's AFB during field carrier landing practice flight testing for the CV variant. Boeing

BOEING
X-32A

Previous page top: The X-32A flies in formation with an F/A-18 Hornet. Previous page bottom and this page: The X-32A seen during various stages of the flight test evaluation. All Boeing

X-32A

BOEING

Previous page top: On 25 January 2001, Boeing successfully completed the X-32A flight tests with a loaded internal weapons bay. Carrying an instrumented Raytheon AIM-120 AMRAAM and a JDAM, the aircraft's weapons bay doors opened and cycled as part of vibration and acoustic testing. Previous page bottom: This almost plan view of the X-32A in flight shows the 2-dimensional thrust-vectoring exhaust nozzle, which was common to both X-32 CDA. Both Boeing

This page above: The X-32B STOVL CDA lift's off from Patuxent River during a short take off run. Boeing

X-32B

Almost aerodynamically identical, much of the data from the X-32A flight test program could be applied directly to the X-32B STOVL flight test program. In July 2000, the Boeing One Team completed a series of tests of its JSF119-614 STOVL propulsion system, during which over 220 transitions between conventional and STOVL operating modes were successfully completed. Up to this time, the STOVL engine had accumulated more than 200 hours of test time, with results matching predictions according to Boeing.

During the tests the system was stable and responsive even when experiencing aggressive pitch, yaw and roll commands. Performance during hover attitude control and stall recovery tests also met or exceeded expectations. Many transitions between vertical and conventional flight modes were accomplished with the JSF119-614 engine running at full power. Boeing met its goal of performing transitions between one and three seconds.

Above: The X-32B lands at Edward's after its first flight from Palmdale. Boeing

Previous page: Prior to entering the flight test program the X-32B conducted engine ground runs in STOVL mode on the Boeing engine run pit at Palmdale on 7 March 2001. Boeing

The X-32B CDA first in-flight conversion from conventional to STOVL mode was conducted on 13 February 2001. Boeing

Engineers at Palmdale installed the flight rated JSF119-614 engine in the X-32B on 12 July 2000, and Boeing conducted a first series of engine runs on 23 September 2000 – four days after the maiden flight of the X-32A CTOL CDA. During the engine runs, the JSF119-614 ran smoothly confirming the team's performance predictions. Engine runs were conducted at low power in order to verify system integrity.

SMI (structural mode interaction) testing of the X-32B was completed on 10 December 2000, during which the aircraft's flight control surfaces were moved at varying frequencies. The SMI evaluation ensured special filters added to the flight control system were functioning properly to prevent vibration in other aircraft components.

A further phase of installed engine runs on the X-32B were completed on 7 January 2001, during which the Boeing JSF One Team conducted flow-switch transitions, with engine thrust re-directed from the aircraft's cruise nozzle to the lift nozzles and back again. The following day, the X-32B successfully completed initial low and medium-speed taxi tests, during which function and integration of crucial aircraft systems were verified. As with the X-32A, the X-32B performed well at 30-knots and again at the medium-speed taxi test at 60-knots

In the build-up to system certification for the JSF119-614, Pratt & Whitney put the JSF119-614 qualification engine through a series of durability tests at its West Palm Beach facility. Accelerated mission tests of the qualification engine were conducted on 15 January 2001. Maximum thrust engine runs in STOVL mode were conducted with the JSF119-614 flight engine installed, over the Boeing engine run pit in Palmdale on 7 March 2001. During the series of engine runs, the test

BOEING
X-32B

Previous page top and bottom: The X-32B was transferred to NAS Patuxent River, Maryland for the final phase of the STOVL flight test program. Patuxent River was more suitable as it was not as elevated as the high desert of southern California. Both Boeing

This page above: For the STOVL phase of the flight test program at Patuxent River, the X-35B lost its distinctive lip intake to aid air ingestion to the engines while in STOVL mode as seen here. Boeing

team repeatedly redirected engine thrust back and forward from the aircraft's cruise nozzle to the lift nozzles.

The last hurdle to clear the JSF119-614 engine for flight certification was final-engine-runs in STOVL mode, which was completed with Pratt & Whitney and Rolls-Royce flight-certified propulsion system hardware and the final validated integrated flight and propulsion control software. During the final engine runs, the test again redirected engine thrust from the aircraft's cruise nozzle to the lift nozzles and back again. By this time, during over 1,300 trials on the STOVL engine test stand, transition times between conventional cruise and vertical thrust and back again had been accomplished consistently in one to three seconds.

Above: The X-32B STOVL CDA taxiing at Edward's AFB. Boeing

X-32B
BOEING

With final flight certification of the STOVL engine the path was clear for the X-32B to begin its flight test program, which commenced on 29 March 2001, when the aircraft successfully completed its first flight, entering a four-month test program to validate the Boeing direct-lift approach to STOVL flight.

Previous page: The Boeing direct lift approach to STOVL flight is demonstrated with the X-32B in the hover at Patuxent River. Boeing

This page: The X-32B edges closer to the ground for a vertical landing at Patuxent River. Both Boeing

During the 50-minute first flight, which was flown in conventional mode, the aircraft lifted off Palmdale and flew to Edward's AFB. Boeing's lead X-32B test pilot Dennis O'Donoghue put the X-32B through a series of initial airworthiness tests, including flying qualities and subsystems checkout. The second flight was conducted on 11 April 2001, with Major Jeff Karnes, USMC at the controls.

The first in-flight conversions from conventional to STOVL mode and back again were conducted during the aircraft's third flight on Friday 13 April, beginning flight test of the direct lift system. O'Donoghue demonstrated the first flow-switch transition, re-directing the X-32B's engine thrust from the cruise nozzle to its lift nozzles and then back again. The transition was accomplished at 180-knots and 9,500-ft during the 58-minute flight.

Later the same day, on the aircraft's fourth flight, O'Donoghue completed seven flow-switch transitions at speeds ranging from 140-185-knots and altitudes between 6,000 and 9,500-ft. The aircraft spent approximately 40 minutes of the 52-minute flight in the STOVL mode. Semi-jet-borne handling quality tests including throttle transients and lift-nozzle thrust vectoring also were completed during the flights.

The X-32B was transferred to NAS Patuxent River, Maryland on 11 May 2001, for the final phase of the flight-test program. Up until then, the aircraft had flown 14 times, accumulating a total of 14.1 flight-hours at Edward's with five different pilots at the controls.

During the 44th flight on 24 June 2001, the X-32B transitioned from wing-borne flight to a hover. The pilot, O'Donoghue, transitioned from fully wing-borne (conventional) to jet-borne (STOVL) flight mode and then smoothly decelerated the X-32B to a steady hover 200-ft above the ground. The pilot then accelerated out of the hover and transitioned back to conventional flight before making a "slow landing." During four other flights the same day, the X-32B completed three additional hovers and numerous transitions to STOVL flight, demonstrating the aircraft's robustness as well as the reliability of the Boeing direct-lift system. In total, the aircraft hovered for eight minutes during the days flying, the single longest sustained hover covering two minutes and 42 seconds. On 27 June, three days after its first hovers, the X-32B completed the program's first vertical landings after the transition from conventional to vertical flight.

On 1 July 2001, the Boeing JSF program successfully accomplished all of its government-defined flight-test requirements, the program reaching the final milestone by making a series of short takeoffs at Patuxent River. Having completed all customer requirements, which included transition to and from vertical flight, hover, vertical landing and short takeoff, Boeing pursued its own objectives before concluding the X-32B flight-test program and on 3 July, Karnes became the first US government test pilot to complete a vertical landing in the X-32B. Karnes performed the landing after transitioning the aircraft to a hover from conventional flight.

Previous page: An elevated view of the X-32B during a vertical landing. Boeing

This page: An underside view of the X-32B with everything hanging out as the STOVL system is engaged for a vertical landing. Boeing

On 5 July 2001, Stone became the first pilot from the UK to complete a vertical landing in the X-32B when during the aircraft's 64th flight he took off conventionally, transitioned to a hover and then landed vertically on the hover pad. On 17 July, the X-32B achieved supersonic flight during both its 67th and 68th flights.

On 28 July, the X-32B completed what was hailed as one of the most successful flight test programs in aviation history, with five flights in one day. On the aircraft's 77th flight, Karnes broke the sound barrier again in the X-32B and demonstrated the range of performance characteristics typical of an operational Boeing JSF: a short takeoff, transition to conventional flight, supersonic flight and transition back to STOVL mode before making a slow landing. The last flight of the day - the 78th and final flight of the X-32B flight-test program - was flown by Royal Navy Lt. Commander Stone and included a series of supersonic runs.

The X-32B CDA validated the Boeing solution to the program's STOVL requirements. Having met all the defined requirements, including a flight in which the aircraft took off in less than 550-ft of roll, Boeing continued performing a series of short takeoffs at successively shorter runway lengths. For the short takeoff, O'Donoghue accelerated the X-32B down the runway to a speed of 80-knots, and then pushed a throttle-mounted flow-switch button that redirected engine thrust from the cruise nozzle to lift nozzles in the underside of the aircraft. The lift nozzles were preset at a position of 60-degrees aft. When the aircraft reached 150-knots and approximately 750-ft altitude, O'Donoghue redirected the thrust back through the cruise nozzle to complete the transition to conventional flight. The X-32B then completed a number of tests before landing.

During flight testing, the X-32B demonstrated more than 100 rapid, one-to-three second transitions to and from STOVL mode, as well as the low pilot workload necessary for operational sorties. From the commencement of flight testing in March 2001, the X-32B demonstrated that the Boeing direct-lift approach to the STOVL requirement was the simplest and lowest-risk solution, according to the manufacturer.

During the JSF program's CDA phase, Boeing clearly accomplished the three main customer objectives by demonstrating: Approximately 90 percent commonality between the two Boeing concept demonstrators, "outstanding" low-speed carrier approach handling qualities and the advantages of its direct-lift STOVL system.

X-32B

X-32B

Previous page top and bottom: The X-32B STOVL variant is show in clean configuration while in conventional flight. Both Boeing

This page: The X-32B STOVL variant is put through its paces. All Boeing

Previous page: The X-32B during take-off. Above: The flight test program demonstrated that the Boeing direct-lift approach to the STOVL requirement was the simplest and lowest-risk solution, but it was not enough to save it from losing the concept demonstration phase of the completion. Top right: The X-32B undergoing ground testing. Bottom right: The X-32B taxis prior to conventional take-off. All Boeing

Although the rival Lockheed Martin X-35 design was selected to go forward for development as an operational aircraft, technologies developed for the X-32 program fed into other programs such as the Boeing F/A-18E/F Super Hornet and the X-45 Uninhabited Combat Air Vehicle demonstrator programs.

With the program over, the two X-32 vehicles were grounded and stored. By 2005, the X-32A had suffered much corrosion damage to the airframe as it had been stored outside in the elements. That year it was transferred to the National Museum of the USAF, in Dayton Ohio to be prepared for display. The X-32B went to the Naval Air Museum at NAS Patuxent River, Maryland in 2005 to be restored for display.

Following a period of outside storage, in 2005 the X-32A and X-32B vehicles were allocated for display at the National Museum of the USAF, in Dayton Ohio and the Naval Air Museum at NAS Patuxent River, Maryland respectively. USAF and USN

Chapter Five

Lockheed Martin X-35 Flight Test Chronology

24 October 2000: The X-35A piloted by Tom Morgenfeld, makes its first flight, from Palmdale, California, to Edwards AFB.
11 November 2000: The X-35A completes its first aerial refuelling.
21 November 2000: The X-35A flies supersonic.
22 November 2000: The X-35A completes it's flight-test program with all objectives and test points achieved.
16 December 2000: The X-35C piloted by Joe Sweeney makes its first flight, from Palmdale, California to Edwards AFB.
31 January 2001: The X-35C flies supersonic.
9-10 February 2001: The X-35C completes a 2,500-mile transcontinental flight from Edwards Air Force Base, California, to Patuxent River Naval Air Station, Maryland. The coast-to-coast flight is believed to be the first ever for an X plane.
1 March 2001: The X-35C, piloted by Lt. Cmdr. Greg Fenton (USN), successfully completes its 100th field carrier landing practice (FCLP) test.
11 March: The X-35C completes it's flight-test program with all objectives achieved.
16 March 2001: X-35B hover-pit testing was completed with full operational thrust achieved.
23 June 2001: The X-35B becomes the first JSF demonstrator to perform a vertical takeoff and vertical landing.
24 June 2001: X-35B achieves its first sustained hover.
3 July 2001: The X-35B flies to Edwards Air Force Base from Lockheed Martin, Palmdale to continue its STOVL flight-envelope expansion.
9 July 2001: The X-35B completes its first airborne transition from STOVL propulsion mode to conventional mode.
16 July 2001: X-35B first vertical landing from wing-borne flight.
26 July 2001: The X-35B performs an automatic short takeoff at 60 knots, using approximately 500-ft of runway. A later flight pushes the aircraft to Mach 1.2.
30 July 2001: The X-35B completes its STOVL flight test program with all objectives achieved or exceeded.
6 August 2001: The X-35B flies from Edwards Air Force Base to Lockheed Martin, Palmdale, bringing the 10-month X-35 flight-test program to a close.

Previous page: The X-35A conducted its maiden flight on 24 October 2000. Above: The X-35B, with everything hanging out conducts hover trials. Both LMTAS

The X-35C could be distinguished from the X-35A/B by its longer span wings. LMTAS

X-35A

First Flight: 24 October 2000
Last Flight as X-35A: 22 November 2000
Total Flights as X-35A: 27
Total Flight Hours as X-35A: 27.4
Maximum Speed achieved: Mach 1.05
g achieved: 5.0
Maximum altitude achieved: 34,000-ft
Angle of Attack: 20-degrees

X-35B

First Flight: 23 June 2001
Last Flight: 6 August 2001
Total Flights: 39
Total Flight Hours: 21.5
Total Vertical Takeoffs: 17
Total Short Takeoffs: 14
Total Short Landings: 6
Total Vertical Landings: 27
Total Mission X Flights: 2
Maximum Speed Achieved: Mach 1.2
g achieved: 5.0
Maximum Altitude achieved: 34,000-ft
Angle of Attack: 20 degrees

X-35C

First Flight: 16 December 2000
Last Flight: 10 March 2001
Total Flights: 73
Total Flight Hours: 58
Total Field carrier landing practice: 252
Maximum Speed Achieved: Mach 1.22
g Achieved: 4.8
Maximum Altitude Achieved: 34,000-ft
Angle of Attack: 20 degrees

Postscript

The latest Boeing F/A-XX concept (above) draws on the companies design experience with previous programs such as the MRF, X-32 Bird of Prey and the X-36 Tailless Agility Flight Research Aircraft (below right). Both Boeing

The design evolution leading to sixth generation fighter technologies is drawing on previous generation aircraft such as the F-22, X-32, X-35, F-35, UCAV and UCAV-N programs. In June 2008, Boeing released a drawing of it sixth generation F/A-XX concept, which appeared to be a twin-engine tailless flying wing design resembling the previous UCAV designs such as the Boeing X-45 and Northrop X-47, but with provision for a pilot. In summer 2009, Boeing released a new drawing of its F/A-XX concept, which appeared to be a twin-engine tailless, blended wing/body design with a more or less conventional type front area incorporating the cockpit and nose. In early summer 2010, Boeing unveiled two separate concepts for a future sixth generation fighter that it had been studying under its F/A-XX program, which at the time the concepts were unveiled had apparently been re-designated NGAD (Next Generation Air Dominance), although this term would be short lived. This concept, it would seem, was aimed at fielding a replacement for the Boeing F/-18E/F Super Hornet strike fighter from around 2025 onwards. In early 2013, Boeing revealed an evolved design of its F/A-XX concept, remaining a twin-engine, tailless design with inhabited and uninhabited variants being considered. The sixth generation concept draws on technology developed for previous programs. The supersonic inlets are diverter-less as on the Lockheed Martin F-35 Lightning II. The most recent concept drawings released show the aircraft with forward canard fore-planes, while earlier drawings of the concept lacked canards.

Under the direction of the Pentagon, DARPA (Defence Advanced Research Projects Agency) is studying sixth generation 'Air Dominance' fighter concepts. In April 2012, the USN issued a RFI (Request for Information) for a potential replacement for the Boeing F/A-18E/F Super Hornet and EA-18G Growler sometime in the 2030's. The main missions for the F/A-XX would be AW (Air Warfare), STW (Strike Warfare) SUW (Surface Warfare) and CAS (Close Air Support). Many more sixth generation fighter program concept designs are likely to appear in the coming years, drawing on the design experience of the major aircraft programs of recent decades. The Boeing concepts clearly draw on linage from a number of programs such as the X-36 Tailless Agility Flight Research Aircraft, X-45 UCAV and earlier programs like the X-32 and its Multi Role Fighter concept forebears. The 2013 F/A-XX is the latest of a run of concepts from the F/X, F/A-XX, F/A-X and back to F/A-XX again. These concepts are being studied in parallel to the UAS (Uninhabited Air System) and UAS+

Glossary

AADC	Allison Advanced Development Company
ACLS	Automatic Carrier Landing System
AFB	Air Force Base
A/F-X	Attack/Fighter Experimental
AGM	Air to Ground Missile
AIM	Airborne Interception Missile
AMRAAM	Advanced Medium Range Air to Air Missile
AMT	Accelerated Mission Testing
ARPA	Advanced Research Projects Agency
ARS	Aerial Refuelling Squadron
ASRAAM	Advanced Short Range Air to Air Missile
ASTOVL	Advanced Short Take-Off and V Vertical Landing
ATA	Advanced Tactical Aircraft
ATF	Advanced Tactical Fighter
ATFLIR	Advanced Targeting Forward Looking Infra-Red
AVT	Advanced Vertical Thrust
A-X	Attack Experimental
B	Bomber
BAE	British Aerospace Systems
BVR	Beyond Visual Range
BUR	Bottom-Up Review
CALF	Common Affordable Lightweight Fighter
CDA	Concept Demonstrator Aircraft
CDDR	Concept Definition and Design Research
CDP	Concept Demonstration Program
CDR	Critical Design Review
CE	Concept Exploration
CEASAR	Component and Engine Structural Assessment Research
CFD	Computational Fluid Dynamics
CTOL	Conventional Take-Off and Landing
CV	Carrier Variant
DARPA	Defence Advanced Research Projects Agency
DERA	Defence Evaluation and Research Agency
DFRC	Dryden Flight Research Centre
DIARS	Distributed Aperture Infra-Red Sensor
DoD	Department of Defence
DSD	Digital Systems Development
E	Electronic
EA	Electronic Attack
ECCM	Electronic Counter Counter Measures
EMD	Engineering Manufacturing and Development

F	Fighter
F/A	Fighter Attack
F/AX	Fighter Attack Experimental
FADEC	Full Authority Digital Engine Control
FCBA	Future Carrier Borne Aircraft
FCLP	Field Carrier Landing Practice
FJTS	Fast Jet Test Squadron
FMS	Full Mission Simulation
FOAS	Future Offensive Air System
FSD	Full Scale Development
FLIR	Forward Looking Infra-Red
GE	General Electric
GR	Ground attack and Reconnaissance
HARM	High Speed Anti-Radiation Missile
HCT	Harrier Conversion Team
HMD	Helmet Mounted Display
HOTAS	Hands On Throttle And Stick
HPC	High Pressure Compressor
HTF	Hybrid Tandem Fan
HUD	Heads-Up Display
ICP	Integrated Core Processor
IHPTET	Integrated High Performance Engine Technology Research
IR	Infra-red
IRIS-T	Infra-Red Imaging System-Tail/Thrust Vector
ISA	Inverse Synthetic Array
JAF	Joint Attack Fighter
JAFE	Joint Advanced Fighter Engine
JASSM	Joint Air to Surface Stand-off Missile
JAST	Joint Advanced Strike Technology
JDAM	Joint Direct Attack Munitions
JDIS	Joint Distributed Information System
JHMCS	Joint Helmet Mounted Cueing System
J/IST	JSF/Integrated Sub-systems Technology
JPO	Joint Strike Fighter Program Office
JSF	Joint Strike Fighter
JSOW	Joint Stand Off Weapon
JTIDS	Joint Tactical Information Distribution System
LOAN	Low Observable Axis symmetric Nozzle
LeRC	Lewis Research Centre
LM	Lockheed Martin
LMTAS	Lockheed Martin Tactical Aircraft Systems
LPT	Low Pressure Turbine
LSPM	Large Scale Powered Model
MATV	Multi-Axis Thrust Vector
MDA	McDonnell Douglas Aircraft
MDC	McDonnell Douglas Corporation

MFDS	Multi-Function Display Screen
MFVT	Mixed Flow Vectored Thrust
MMD	Moving Map Display
MoD	Ministry of Defence
MoU	Memorandum of Understanding
MRF	Multi Role Fighter
N	Naval
NASA	National Aeronautics and Space Administration
NATF	Naval Advanced Tactical Fighter
NG	Northrop Grumman
NGAD	Next Generation Air Dominance
OARF	Outdoor Aerodynamic Research Facility
OSA	Open Systems Architecture
P&W	Pratt & Whitney
PHM	Prognostic Health Management
PST	Pacific Standard Time
PWSC	Preferred Weapon System Concept
RAF	Royal Air Force
RALS	Remote Augmented Lift System
RFP	Request For Proposal
RN	Royal Navy
RR	Rolls Royce
SAR	Synthetic Aperture Radar
SBAC	Society of British Aerospace Companies
SDD	System Development and Demonstration
SLAM	Stand-off Land Attack Missile
SLAM ER	Stand-off Land Attack Missile Enhanced Response
STO	Short Take Off
STOL	Short Take-Off and Landing
STOVL	Short Take-Off and Vertical Landing
USA	United States of America
USAF	United States Air Force
USMC	United States Marine Corp
USN	United States Navy
UK	United Kingdom
VAAC	Vectored-thrust Aircraft Advanced Control
VISTA	Variable Stability In-flight Simulator
VSTOL	Vertical/Short-Take-Off and Landing
X	Experimental

Centurion Publishing

ISBN: 978-1-903630-20-4